FEMINISM IN
TWENTIETH-CENTURY
SCIENCE, TECHNOLOGY,
AND MEDICINE

WOMEN IN CULTURE AND SOCIETY

A SERIES EDITED BY CATHARINE R. STIMPSON

FEMINISM

IN TWENTIETH-CENTURY

SCIENCE, TECHNOLOGY,

AND MEDICINE

EDITED BY

ANGELA N. H. CREAGER

ELIZABETH LUNBECK

& LONDA SCHIEBINGER

THE UNIVERSITY OF CHICAGO PRESS • CHICAGO AND LONDON

ANGELA N. H. CREAGER is associate professor in the Department of History and the Program in History of Science at Princeton University. She is the author of *The Life of a Virus: Tobacco Mosaic Virus as an Experimental Model, 1930–1965,* published by the University of Chicago Press. ELIZABETH LUNBECK is associate professor of history at Princeton University. She is the author of *The Psychiatric Persuasion: Knowledge, Gender, and Power in Modern America.* LONDA SCHIEBINGER is the Edwin E. Sparks Professor of History of Science at Pennsylvania State University. She is the author of *The Mind Has No Sex? Women in the Origins of Modern Science; Nature's Body: Gender in the Making of Modern Science;* and *Has Feminism Changed Science?* and is the editor of *Feminism and the Body.*

The University of Chicago Press, Chicago 60637
The University of Chicago Press, Ltd., London
© 2001 by The University of Chicago
All rights reserved. Published 2001
Printed in the United States of America
10 09 08 07 06 05 04 03 02 01   1 2 3 4 5

ISBN: 0-226-12023-6 (cloth)
ISBN: 0-226-12024-4 (paper)

Library of Congress Cataloging-in-Publication Data

Feminism in twentieth-century science, technology, and medicine / edited by Angela N. H. Creager, Elizabeth Lunbeck & Londa Schiebinger.
    p. cm. — (Women in culture and society)
Includes bibliographical references and index.
ISBN 0-226-12023-6 (alk. paper) — ISBN 0-226-12024-4 (pbk. : alk. paper)
    1. Feminism and science. I. Creager, Angela N. H. II. Lunbeck, Elizabeth.
III. Schiebinger, Londa L. IV. Series.

Q130 .F46 2001
500'.82—dc21

                                                                    2001027410

CONTENTS

MEDICINE

FOREWORD

In 1964, C. H. Waddington, a distinguished biologist who was sympathetic to women scientists, wrote to the equally distinguished biologist Theodosius Dobzhansky. Waddington wanted Dobzhansky to be less critical of a woman geneticist. "I think," Waddington wrote, "that women biologists in America have in any case a very difficult job to get themselves accepted—look at Barbara McClintock, as the most extreme case of an absolutely first-rate person who has been forced into the position of an eccentric recluse; and the positions of Salome Waelsch, Jane Oppenheimer, Dorothea Rudnick is only slightly better." The name Jane Oppenheimer struck me with special force, for I had gone to a women's college where we gossiped endlessly about all our professors and talked seriously about a few. Among them was Jane Oppenheimer, who must then have been in mid-career. She provoked awe in the pre-med students whom I knew, and they were terrified of displeasing her. Science majors and nonscience majors alike knew Miss Oppenheimer to be one of the most notable members of our faculty and a formidable symbol of what women might be and do. Reading of Waddington's letter, I learned that the tall woman who had worked in her lab and classroom with such apparent power and influence might have been—through no fault of her own—far less powerful and influential in her profession than she ought to have been. To be sure, I was miffed that Waddington did not evaluate the status of scientists with tenure at my college more highly; but, more significantly, his letter provoked the emotions that tales from women's history so often do: anger at the treatment of women, grief at the diminishment of their talents, admiration for their resilience, courage, and persistence.

Waddington's letter is archived in the Dobzhansky Papers at the

American Philosophical Society. I discovered it in a note in a remarkable essay by Scott F. Gilbert and Karen A. Rader in this volume, *Feminism in Twentieth-Century Science, Technology, and Medicine,* a landmark collection edited with imagination and skill by three vital figures in the study of women and gender in science: Angela N. H. Creager, Elizabeth Lunbeck, and Londa Schiebinger. Because of the centrality of science, technology, and medicine to modern life, this book ought to be read by scientists and nonscientists alike. As a nonscientist, I am grateful for its ability to clarify sophisticated questions without condescension or oversimplification. *Feminism in Twentieth-Century Science, Technology, and Medicine* builds scrupulously on previous work about women, gender, and science, but its focus is fresh. What, it asks, has been the influence of feminism on women and science? How can scholars make visible and "reveal the potent . . . transformations effected by feminism within institutions, professions, and disciplines" of science (p. 4)?

The meanings of feminism are famously varied, but the editors write plausibly that "bottom line" feminism recognizes the importance of gender in history, culture, and society but opposes the destructive consequences for women that gender systems can breed—the rigid roles, the inequities, the violence. "Bottom line" feminism connects gender to the other systems that structure our lives and individual identities. These essays reflect such connections. Evelynn M. Hammonds's piece about the HIV/AIDS epidemic in the United States shows how interwoven gender, sexuality, and race have been in the responses to AIDS. The story of Salome Waelsch, whom Waddington mentions, is a case study of the links between misogyny and anti-Semitism. In 1932, when Hitler was coming to power, she was as yet unmarried, a young biologist who had just completed her Ph.D. at the University of Freiburg. She had worked under an embryologist who was to win the Nobel Prize in 1935 but who was reluctant to work with her. Waelsch found him "old and an anti-Semite, and also a strong anti-feminist to participants in his experiments" (p. 80). Looking for postdoctoral work in Berlin, she was rejected with "You, a woman and a Jew—forget it!" (p. 82).

After over three decades of toil, feminism has made enormous contributions to our understanding of science, technology, and medicine. Feminist writers have, with varying degrees of politeness, shown that "scientific authority has not been gender neutral" (p. 1); that technological innovations have been credited to and benefited men more than women; and that medical narratives and procedures have kept gender stereotypes and inequalities alive. The treatment of women in science has helped to permit these huge errors. Again and again, women have bumped up against a silicon ceiling as they have sought to advance in their professions. They have been consigned to fields such as X-ray crystallography or embryology after the 1920s rise of classical genetics, fields "considered peripheral, full of material details, and full of women practitioners" (p. 78). They have made their living less often in research universities than in private foundations or institutions, teaching colleges, or, like Jane Oppenheimer, in women's colleges.

Going beyond critique, feminist analysis brings out the ways in which "women make [the world] work" (p. 182). In so doing, it profoundly alters intellectual and academic practices. After reading the work of Evelyn Fox Keller, whose publications include a biography of Barbara McClintock, one cannot imagine thinking about the history and philosophy of science without thinking about gender. After reading Ruth Schwartz Cowan, who pioneered the linkage of women's studies and the history of technology, one cannot imagine thinking about the history of technology without thinking about gender and deconstructing technology's masculine codes. When this field includes women as consumers, as Cowan argues that it must, women are as much the designers of systems and machines as men.

As a social and political movement, feminism has been a constructive force in efforts to reform the academy and science. In the 1960s and 1970s, measures were adopted to promote equal educational opportunities. Between 1990 and 1994, twenty-five pieces of legislation supported the feminist revolution in health, including the establishment of the National Institutes of Health Office of Research on Women's Health and the requirement that clinical trials include women. None of this would have happened without a women's movement. As Evelyn Fox Keller points out in her essay "Making a Difference: Feminist Movement and Feminist Critiques of Science," feminism as a social and political movement preceded and was a matrix of feminist scholarship. Yet even people who have benefited from the contributions of feminism and feminist scholars disdain the F-word. Their anxiety marginalizes and effaces feminist accomplishments. Feminism, the editors suggest, is feared because it is perceived as bringing an agenda and concerns from outside science into science. The possibility of being both a feminist and a scholar or researcher is dismissed out of hand. In brief, feminism may be the victim of a double standard, covertly used but overtly disdained as a hijacker of scientific virtue.

This volume unravels such complexities with poise, balance, knowledge, and energy. Although the contributors show us past errors about women and gender, they are too realistic about the processes of inquiry to believe that feminists have the last, wholly accurate word. Their essays stimulate our explorations rather than bringing closure to them. One of this volume's many strengths is its comprehensiveness. It takes up not one but several important disciplines and activities, ranging from archaeology to computer science to the new reproductive technologies. Of these, engineering has perhaps been the most untouched by feminism; primatology has been among the most receptive. The editors ask us to see these disciplines for themselves and as activities that often touch and mirror each other. One consistent theme is the feminist conviction of the necessity of seeing multiple forms of *interactivity* among the constituent parts of reality—be they the egg and the sperm in the traditional model of human reproduction, females and males in primate groups, software programmers and users of the personal computer, or the ways in which we name phenomena and how we experience them. To see life as interactive is to destroy—cheerfully and irrevocably—the tired old binary association of the male with activity and the female with passivity.

At their finest, science, technology, and medicine inspire in us a sense of wonder—science because of the beauty and strength of the natural forces and patterns it explores; technology because of the value and robustness of the systems and tools that it builds and distributes; and medicine because of its ability to enhance life and to heal some of our most savage illnesses and wounds. One of her students told me that Jane Oppenheimer was demanding in the embryology laboratory. Experiments had to be done rigorously and precisely. And yet, this student added, the lab made nature seem exciting, an unending source of mysteries to be explored. *Feminism in Twentieth-Century Science, Technology, and Medicine* is carefully researched and written, but it should engender in its readers a sense of wonder about the achievements of a great intellectual and social movement, and even more broadly, it should renew readers' amazement at the capacities of the ancient yet ageless human activities of creation, exploration, invention, and care.

Catharine R. Stimpson

ACKNOWLEDGMENTS

This volume emerged from a workshop, "Science, Medicine, and Technology in the Twentieth Century: What Difference Has Feminism Made?" held in October 1998 at Princeton University and organized by the Women's Caucus of the History of Science Society. This workshop was conceived as a stepping stone between the large national conference on women, gender, and science held at the University of Minnesota in 1995 and the conference on women and gender in science, medicine, and technology held at St. Louis University in October 2000. The Princeton workshop was specifically focused on feminism itself and on the diverse changes it has generated in disciplines of science, technology, and medicine over the course of the twentieth century. Bringing feminist historians of these sometimes disparate fields into a common conversation, the workshop was the occasion for grounding theoretical concerns in specific case studies. This volume showcases current innovative scholarship while at the same time inviting further engagement among feminists working on science, technology, and medicine.

Many people contributed to the success of the workshop and ensuing volume. In 1997, the History of Science Society supported a planning meeting involving members of its Women's Caucus interested in organizing a new scholarly event on gender and science. A coterie of grant writers—led by Arleen Tuchman and including Karen Rader and Pamela Mack—emerged from that planning meeting. We particularly appreciate their efforts, which were critical to the resulting workshop. We also thank Sally Kohlstedt, Rima Apple, Margaret Rossiter, Evelynn Hammonds, and Peggy Kidwell for their sage advice on organizational strategy and style. From the beginning, Norton Wise supported the workshop at Princeton and we are grateful for his involvement. Carolyn Goldstein, Arwen Mohun, Naomi Oreskes, Andrea Rusnock, and Abha Sur made important contributions to discussions at the workshop toward planning a subsequent and larger event. We are pleased that Charlotte Borst at St. Louis University joined with these historians (and others) to make the 2000 conference a reality.

We gratefully acknowledge funding for the workshop provided by various offices and departments at Princeton University: the Office of the President, Department of History, Program in History of Science, Program in Women's Studies, Council on Science and Technology, Humanities Council, and the Shelby Cullom Davis Center for Historical Studies. The National Science Foundation also provided major funding for the event through SBE grant number 98-06230.

Susan Bielstein, our editor at University of Chicago Press, imparted valuable support and guidance along the way. We are also thankful for the helpful suggestions provided by the anonymous referees who reviewed the book in various stages. Gail Schmitt and Suman Seth, graduate students at Princeton, provided much-appreciated assistance in preparing the manuscript for press. We are grateful to Pamela Bruton for her fine copyediting. Lastly we thank our partners and sons for their support and patience throughout our labors on the workshop and volume.

ONE

Introduction

ANGELA N. H. CREAGER, ELIZABETH LUNBECK,
AND LONDA SCHIEBINGER

It is now a commonplace that changes in the nature and reach of science, technology, and medicine have characterized social, political,
and material developments of the twentieth century. During the past
twenty-five years, feminist scholars have scrutinized developments in
each of these domains, pointing out the many ways in which scientific
authority has not been gender neutral, in which technological innovations have conventionally been attributed to—and benefited—men
more than women, and in which medical representations and procedures have perpetuated gender stereotypes and inequalities. Moreover,
the exclusion of women from participation in each of these domains
has been abundantly documented. The extensive and rich literature on
these issues has been implicitly organized around a critical program,
a literature in which it is proposed that feminism can provide a qualita-
tively better lens through which to view historical processes. Yet this
critical program has been somewhat fragmented, with the large body
of scholarship on gender and science isolated from that on gender and
technology, which in turn has been formulated separately from feminist studies of medicine. Although some scholars have crossed these
disciplinary divides, separate feminist programs have emerged in each
of these three domains.[1]

This volume seeks to advance this active body of scholarship in
three ways: First, we have brought together discussions of science,
technology, and medicine while attempting to maintain an appreciation for the distinct ways in which scholars working on the history of
these areas have approached questions about women and gender. Because these domains are increasingly considered together, for
the purposes of synthetic scholarship, communications, and graduate

1

study, we wish to explore the comparisons invited by this grouping.[2] By restricting our attention to twentieth-century developments, we have sought to enhance specific historical comparisons across these three areas. Second, we have attempted to shift the focus from women and gender to an emphasis on *feminism,* examining it in its diverse instantiations as a source for historical change during the past century. Many scholars have used feminism as a theoretical framework through which to understand and critique science, technology, and medicine. Yet feminism itself has been a part of twentieth-century history and has coexisted, however uneasily, with developments in each of these areas. Finally, we have sought to bring these connections into focus by asking, What difference has feminism made to science, technology, and medicine? How have feminists contributed to the intellectual ferment of the past quarter century? Critique, philosopher Alison Wylie has urged, involves the systematic restudy of particular questions and often results in new approaches. The chapters in this volume thus approach feminists and their questions as a source of innovation and transformation in various scientific, technological, and medical developments.

Feminists have enjoyed success in their efforts to open many fields to women as participants, a subject that has been extensively studied by both scholars and government agencies.[3] Professional opportunities for women, however, have been highly uneven; some scientific disciplines (e.g., laboratory biology) and medical specialties (pediatrics) more readily welcome women than others (physics, cardiology).[4] Looking to professions associated with technological research and development, there are few fields of engineering where women participate at equal levels and at equal status with men. But the effects of feminism have not been restricted to altering employment and professional opportunities for women. In many areas, feminism has been the impetus for fundamentally changing the theoretical underpinnings and practices of research. The chapters in this volume provide several examples of these transformations. In primatology today, nonhuman females are no longer passed off as docile creatures who trade sex and reproduction for protection and food but are instead studied for their own unique contributions to primate society. In archeology, spearthrowers found in women's graves are no longer immediately considered purely ceremonial devices or evidence testifying to the transfer of property but are sometimes interpreted as evidence of women's hunting activities or toolmaking prowess. In medicine, clinical trials now routinely take into account physiological differences between women and men. In computer science, feminist epistemologies have been rec-

ognized by some practitioners as frameworks for rethinking the practice of programming and the human-interface problems associated
with the growing reliance on computers.

At another level, the chapters in this volume emphasize that feminism has been important historiographically, changing the way that
science, technology, and medicine are understood, necessitating a re-
drawing of the boundaries for these domains. Historians of technology
stress that technology includes diffusion and consumption as well as
design and production, and that since the beginning of the century
women have had profound effects as consumers and on the reshaping of technological artifacts, particularly those reaching the domestic
sphere. Furthermore, women's movements have played an important
role in contributing to government regulation of industry, through ad-
vocating laws governing occupational health and safety as well as consumer safety. Thus the history of technology encompasses the changing practices of consumption and women's political activism as well
as activities within manufacturing sites and engineering offices. In the
history of medicine, attention to a range of women health workers and
to the experiences of patients has expanded the scope of feminism's
impact beyond the changing number of women physicians and gendered conceptions of disease. Feminism has also been historically intertwined in complex ways with specific medical developments, such as
the availability of prenatal diagnosis and the changing normalization
of emotion in psychiatry and psychology.

In focusing on the ways feminism has conceptually reordered the
fields of science, technology, and medicine, we do not wish to lose
sight of the significance of women's political movements in relation to
scholarship. It would be highly contradictory to imagine that *feminism*
could flourish in the academy without women. The women's movement in the nineteenth century propelled women into European and
North American universities in the 1880s and 1890s. The story of how
American universities opened graduate education to women in the
1920s and 1930s has been masterfully detailed in Margaret Rossiter's
two volumes, *Women Scientists in America*.[5] While we do not directly
discuss the political struggles to pass major legislation guaranteeing
women rights—equal education, pay, and opportunity (e.g., the Equal
Pay Act of 1963; Title IX of the Education Act Amendments of 1972;
the Equal Opportunity Employment Act of 1972)—these developments underlaid the growing strength of women in the professions and
provided a groundwork for feminist innovations that shape how science, technology, and medicine are practiced. Similarly, behind the
accounts included here of biomedical changes—in birth control and

in the understanding of mental health and AIDS—stands the historic enactment by the U.S. Congress of legislation that supported the feminist revolution in women's health care, in particular through the creation of the National Institutes of Health Office of Research on Women's Health in 1990 and the Women's Health Initiative in 1991. Between 1990 and 1994, Congress enacted no fewer than twenty-five pieces of legislation to improve the health of American women, ranging from requiring that females be included in clinical trials to establishing new federal regulations for mammography.[6]

Attention to women's movements and their political achievements, however, has tended to reinforce a presumption that feminism stands outside mainstream institutions and that change results from critique-from-without. The examples of change offered in this volume, by contrast, reveal the potent, but less visible, transformations effected by feminism within institutions, professions, and disciplines. As the feminist political scientist Mary Katzenstein has observed, the past two decades appear as a quiescent period for feminism when one looks for political activism as a barometer of social change. However, her examination of the impact of feminism-from-within on two conventionally male-dominated institutions, the Catholic church and the U.S. military, shows the continuing vitality of feminism as a source for profound change.[7] Similarly, accounts provided here of changes in developmental biology, primatology, archeology, and psychiatry, among the many fields examined, illustrate how feminism has been assimilated, and indeed at times "domesticated," by professionals in their own disciplines. This process of change-from-within has sometimes had the paradoxical effect of effacing feminism, as some practitioners shy from attributing to ideology or politics what now appear as merely reasonable changes in agenda and methods. Part of our project, then, is to draw attention to the many unacknowledged but profound effects of feminism in science, technology, and medicine.

SCIENCE

Our first section deals with science; the chapters presented here—by a philosopher, a primatologist, a biologist, and two historians of science—underscore the collaborative and interdisciplinary character of gender studies of science and testify to the kind of intellectual innovation that can emerge from active collaboration between scientists and humanists. These chapters also demonstrate that gender analysis of science is not altogether external to science; rather, it also has been actively cultivated by working scientists who have not only brought feminist insights to bear on their work but have also contributed to key aspects of feminist theory and practice. For example, some primatologists themselves, not academic outsiders, have provocatively claimed that primatology is a feminist science.[8] The chapters in this section identify and analyze concrete examples of feminist practices in specific sciences. We have not here addressed the question of how women have changed science. Nor have we tried to define or identify a singular "feminist science." Rather, we are interested in examining the categories of gender analysis that enable feminist contributions to mainstream scientific thinking. These methods, which Londa Schiebinger has elsewhere called "tools of gender analysis," are intentionally open-ended and flexible, as diverse as the variants of feminism and of science.[9] These tools are not always peculiar to feminist analysis but include good history, sharp critical thinking, good biology, and precise use of language—analytics that feminists are applying to answer new questions. Furthermore, some tools of gender analysis transfer easily from science to science, while others do not. Most important for our purposes, these categories of analysis are not presented here in an abstract fashion only. Each author has provided case studies and examples of what new knowledge these analytics have produced.

Archeology has not traditionally been part of the gender studies of science; its inclusion is long overdue. Alison Wylie's chapter leads off with a discussion of feminist revisions in archeology. As Wylie documents, gender analytics came late to archeology—around 1984—despite feminism's strong showing in the neighboring fields of history and anthropology and despite what may have appeared to be the relevance of gender analysis to central issues in archeological research, such as sexual divisions of labor, the significance of gender symbolism, and the gendered use of tools. As a measure of the "newness" of feminism to the discipline, Wylie can enumerate the key conferences and volumes on gender and archeology in the space of approximately one page. Exploring the process by which archeology opened up to gender studies and the substantial changes brought about since the mid-1980s by "the archeology of gender," Wylie identifies two particular critical techniques common in feminist work in this field. The first, "countering erasures," invokes the familiar notion of making women and their activities visible. Seemingly simple conceptually, this analytical tool has thrown into question the pride of place given to hunting practices and toolmaking activities traditionally associated with males. A second technique of feminist scholars has been "countering stereotypes" concerning the capacities and roles proper to women and men.

Wylie gives the example of the practice of sexing prehistoric skeletons according to present-day assumptions concerning sexual dimorphism. Archeologists have long used "robustness" to sex the skeletal record of Australian Aboriginals, which has resulted in more bones being identified "male" than could reasonably have been the case. Wylie also analyzes a number of fruitful feminist theories that are bringing into view aspects of prehistoric households and communities that simply have not been investigated in the past.

If Alison Wylie's chapter asks why feminism came late to archeology, primatologist Linda Marie Fedigan's chapter asks why feminist perspectives have been so successful in primatology that the study of nonhuman primates is now often celebrated as a "feminist science." Yet, at the same time that primatology has become the "darling child" of feminist scholars, most primatologists vehemently deny that theirs is a feminist science. Fedigan's chapter thus explores a paradox also highlighted by Wylie: given that primatology so quickly self-corrected in response to feminist critique, why are so many practitioners who use feminist methods loath to call themselves feminist?[10] Why practice feminist virtues while mistrusting the label "feminism"?

Fedigan's chapter details in valuable ways the many feminist insights that have contributed to primatology. Revising theory, some primatologists have reconceptualized power relations among baboons, for example, to show that it is not those using brute force who can mobilize allies but those who actively build systems of reciprocal friendships. Also at the theoretical level, other primatologists have undermined the notion that male primates "own" females by showing how in some primate societies females first distribute themselves according to the available food and water, and males then distribute themselves according to the pattern of available females. At the level of language use, by rejecting gender-loaded words like "harem," primatologists have been able to "see" that in many primate societies a network of related females, not males, forms the stable core. In this chapter Fedigan moves away from her earlier work that defined and documented ways in which primatology has become a feminist science toward a thoroughgoing implementation of "tools of gender analysis" in order to show how—in very different ways and at different epistemic levels—primatologists are incorporating feminist analyses into their work.

Developmental biologist Scott Gilbert and historian of science Karen Rader discuss the extraordinary success women and feminism have enjoyed in developmental biology, one measure of which is the fact that over a third of recent presidents of the Society for Developmental Biology have been women. Women's involvement in developmental biology, in contrast to archeology, stretches back to the 1930s. Drawing on oral histories and collective biographies of women biologists, Gilbert and Rader explore the strange brew of social and intellectual currents that brought women into the field. They also document how old conceptions were challenged and transformed. A 1976 brochure published by the Women's Caucus of the Society for Developmental Biology, entitled *Sexisms Satirized,* enjoined developmental biologists to remove sexist language and "reassess their objectivity in future publications." Methodologically simple, this type of corrective led to deeper revisions in research programs investigating sex determination, fertilization, and brain development. Many of these results have been mainstreamed in textbooks in the field and accepted as "good" science, not necessarily as feminist science.

In her chapter, Evelyn Fox Keller offers an analytic map to concepts in gender studies of science and some reflections on the past twenty-five years of work in this area, arguing that success has brought a depoliticization of the field. As she sees it, as feminism has become more successful over the past two decades, scholars have become more "academic," focused more on the growth of knowledge—and perhaps also on their place in the academy—than on broad political goals. In contrast to the authors of the other chapters in this volume, she cautions feminist academics against assuming too much credit for their effectiveness in changing science and other realms of knowledge. Keller argues that the "heroine" of feminist reforms is the broader women's movement—and the complex of social changes that it brought—and not feminist scholarship per se. Feminist scholarship, she contends, is but one of the many by-products of the women's (and men's) movement that changed fundamentally the meaning and practice of gender in American culture.

Keller offers two examples of the fruits of feminist investigations in biology. The first is the now familiar example of the story of fertilization, how the narrative of an active, forceful, and self-propelled sperm that fertilizes a passive egg has gradually been rewritten. Her point is to show how language can shape the thinking of working scientists. In this case, feminism entered by "waking up," in Emily Martin's words, the metaphors with which scientists think. Considering the egg as possibly coactive allowed scientists to recognize its contributions to fertilization.[11] Keller's second example concerns the revisions in thinking about the role of "maternal effects" whereby the biological contributions from females to their developing offspring are now credited with a larger evolutionary significance than previously recognized.

A central question in gender studies of science is, Why has feminism not enjoyed uniform success across the sciences? We have not included a paper on the physical sciences because, to date, it has not been shown that feminism has deeply touched the field. Unlike the case of biology, where the conceptual assimilation of feminism has been mirrored by increasing numbers of women practitioners, no equally dramatic increase in participation has marked physics. Nor are there many documented instances of change in the theories and practices of physics.[12] It is, nonetheless, premature to conclude, as do some, that there simply are no gender dimensions to uncover in physics.[13] Rather, the question of gender in physics remains open—a question requiring sustained and careful research.[14] The number of scholars currently trained in both physics *and* gender studies is very small. Signs are encouraging as scholars, such as Karen Barad and Amy Bug, undertake work in this field.

TECHNOLOGY

As both Carroll Pursell and Ruth Oldenziel remind us in their chapters, the field of the history of technology emerged after World War II from the culture and interests of engineers, reflecting the priority they gave to invention and design. To the extent that professional engineering has been male dominated in practice and masculinist in ideology, it is unsurprising that conventional histories of technology, even those focusing on the twentieth century, have paid little attention to feminism. The five chapters gathered in this section reconsider the relevance of feminism to the history of technology by demonstrating both how feminist approaches to history have revealed the earlier involvement of women and women's movements in technological change and how technological developments of the last two decades have interacted with and responded to feminism.[15]

Pursell provides an insightful participant's perspective on two decades of growing interest in feminist questions and methodologies among historians of technology. In the 1970s, feminist inquiry produced mainly recuperative history, showing that women, too, had been engineers and inventors, patent holders and technological entrepreneurs.[16] Ruth Schwartz Cowan's 1976 paper at the annual Society for the History of Technology meeting, addressing whether women's experience of technology had been historically different from that of men, served as a catalyst to extend the purview of feminist history of technology into women's history more generally.[17] Subsequently, in the wake of Judith McGaw's pointed observation that the field lacked a gender-conscious history of men, the social construction of masculin-ity has emerged as a prominent theme in technology's history.[18] Indeed, Pursell himself has been a major contributor to this development, showing how magazines such as *Popular Mechanics* promoted the benefits of "boy engineering" and unambiguously coded technology masculine.[19] As he observes, attention to the presumed masculine agents of technology has also brought into view other ideological di-mensions, particularly as related to race and class. Pursell argues that feminist and race theories are important and fruitful tools for histori-ans of technology, as well as for those committed to opening access to technology in contemporary society.

This increasing interest in the gendered dimensions of technology that Pursell recounts has been part of a larger methodological reassess-ment of the history of technology, in which users as well as producers have been considered agents in the process of change. Oldenziel pro-vides a keen analytic perspective on this important historiographical transition, pointing to Ruth Schwartz Cowan's classic essay "The Con-sumption Junction."[20] By expanding the historian's range of relevant technological changes to include use as well as design, consumption as well as production, tacit knowledge as well as patent activity, a fuller range of historical actors has come into view, including women, workers, and ethnic minorities. Indeed, as Oldenziel shows, the mascu-linization of technology was itself a historical process, with women resisting the process all along the way.

This broader historical approach illuminates the ideological role of women's movements as well as of the activities of women in shaping technology. Early-twentieth-century feminists took seriously the politi-cal responsibilities of their roles as household managers and extended the sphere of women's activities to include municipal housekeeping as well. Manufacturers were aware of the importance of women's groups in shaping patterns of consumption, and utility companies sought out women professionals to ease the acceptance of electricity and the new appliances it powered—electric ranges, washing machines, and vac-uum cleaners—into homes.[21] Oldenziel argues that this midcentury al-liance between manufacturers, home economists, and activists con-tributed to consolidating the myth of the middle-class housewife that Betty Friedan subsequently critiqued in *The Feminine Mystique*. Thus, in Oldenziel's view, the 1960s debates over sex roles called into ques-tion not only patriarchal structures but also earlier articulations of feminism and their accommodations to technology.

Pamela Mack's chapter reveals that even the classic focus of histori-ans of technology—engineers and engineering—offers new historical insights when examined in concert with the history of U.S. feminism in its many forms. On the one hand, equal-rights feminists have sought throughout the century to challenge barriers to the participation of women as professional engineers, with increasing visibility since the 1960s (although the number of women engineers remains modest in many fields). Difference feminists, on the other hand, have challenged more fundamental aspects of engineering. Earlier in the century, many difference feminists worked out of the women's reform movement as activists. These women, Mack argues, played a crucial role in changing the regulatory environment for industry, which, in turn, had far-reaching consequences for engineering itself. These reform activities also opened up career opportunities for women in new technical fields, such as industrial medicine. The participation of women in engineering fields increased substantially beginning in the 1970s, but in contrast to the pre–World War II movements, neither the growing equal-rights feminism nor the more radical feminist critiques of industrialization and technology have had much impact on the practices of postwar engineering. This may reflect the heterogeneity of feminism itself. For instance, practicing women engineers, who likely advocate equal rights in their professional lives, are unlikely to share a critical view of technology with, say, ecofeminists.

The other two chapters in this section focus on the historical significance of feminism for two specific areas of rapid postwar development: computers and prenatal diagnosis. Michael Mahoney offers an analysis of computers as a case study in the complex relationship between gender and technology. He takes the masculinization of computing technology as a historical problem rather than a given, particularly in view of the dominance of women in the field of computing up through the early development of computing machines in the 1940s. In his view, the masculinization of software engineering has taken a path different from that of hardware. The culture of programmers and software engineers, whether corporate employees or hackers, is characterized by a collective style of work that some scholars have viewed as highly compatible with a feminist approach.[22] While Mahoney finds some problems with this argument, he views the contemporary situation in computer software design as potentially very responsive to feminism. The engineering model for program development, in which technical problems have been given highest priority, has in many respects failed. Currently there is a perceived need in software engineering for designers who can carefully attend to the needs of users and who can make visible the kinds of devalued and ordinary tasks that go into getting work done in real-world situations. To the degree that feminist analysis has helped make clear the process of privilege and devaluation that often accompanies the gendering of objects and processes, it has the potential to provide critical new ways of thinking in the lucrative world of software development. In response to Mack's conclusion that feminism has had little impact on the practice of engineering in the past two decades, Mahoney argues that contemporary computer technology might prove to be an exception.

Looking toward the realm of medical technologies, Ruth Schwartz Cowan focuses her attention on the history of prenatal diagnosis, a case where the pertinence of feminism is undeniable.[23] Following her own methodological principle (of focusing on the "consumption junction"), Cowan has approached the system of prenatal diagnosis from the perspective of the patients—who in this case are necessarily women. Viewed from this consumer's perspective, the technological system in which prenatal diagnosis is embedded includes abortion services as well as techniques such as amniocentesis. The contested political history of abortion has thus been intertwined with the diffusion of prenatal diagnosis, slowing both publication and provision of services by physicians. By rallying for the legalization of abortion, the women's movement helped advance the cause of prenatal diagnosis as well. However, to the degree that prenatal diagnosis has entailed a greater medicalization of pregnancy, it is also counter to some aims of the women's movement. Even more troubling is the use of prenatal diagnosis for sex selection, which raises ethical dilemmas unanticipated by earlier feminists. Thus the case of prenatal testing shows the paradoxes that have emerged as feminism interacts with technology: feminism has both accelerated and strongly critiqued the development of these new medical technologies. Cowan speaks against the point of view of some feminist critics, urging that a historical assessment of these technologies does not support claims that they have led to the greater subordination of women. Yet she recognizes the irony that feminists appear on both sides of the debate, both advancing and resisting new regimes of prenatal medical intervention.

Taken together, these chapters bring into view the altered trajectories and insights that feminism has contributed to technological developments. These are somewhat distinct from those illuminated in the chapters on science. Feminist critiques of science have often had a scientific edge, revealing sources of bias and prejudice, as well as calling attention to new interpretations. Regardless of whether or not scientists embrace feminist politics, they appreciate critical insights that will help them achieve better accounts of nature. For this reason, some scientists have assimilated—at least in part—feminist approaches. By contrast, as Oldenziel points out, feminist critiques of technology have focused on expanding the boundaries of the field to acknowledge the role of consumers, who have often been women, as well as those of designers and producers, who have often been men. Even as the old dichotomy between men as producers and women as consumers is breaking down, the role of consumers in shaping artifacts means that users play an important role in directing technologies. As Oldenziel observes, "Technologies—perhaps more so than scientific practices—are to a large degree constituted through their use."[24] Thus what is at stake in the feminist interventions in technology is not scientific truth or unbiased medical representation (discussed below). And in fact, feminists might shape the development of technological artifacts and systems most profoundly, not as critics, but as politically aware consumers.

MEDICINE

Analyses of medicine and the gendered body have long figured centrally in feminist scholarship. A narrative that told of women's exclusion from the medical profession and of abuse at the hands of male physicians was central to feminism both in the academy and beyond in the 1970s. A large body of scholarly work on the medical, surgical, and psychiatric misrepresentation and mistreatment of women's ailments was echoed on a more popular level in books such as *For Her Own Good,* by Barbara Ehrenreich and Deirdre English,[25] and the enormously influential *Our Bodies, Ourselves,* which appeared in 1973.[26] The task, as many saw it, was to set the record straight, to envision what a more accurate historical and contemporary account of women's bodies, healthy and diseased, might look like. This impulse to get the story right, to peel away the distortions imposed by male physicians and commentators, proved enormously productive, fueling much of the work in the growing field. Carroll Smith-Rosenberg, for example, asked in 1972 whether nineteenth-century women were "really" hysterical or was their putative hysteria better seen as an understandable reaction in a society that offered them no power and no voice.[27] More recently, Emily Martin's *The Woman in the Body* showed how medical textbooks unwittingly incorporated gender into the purportedly neutral and scientific descriptions they offered of menstruation, fertility, reproduction, and childbirth.[28] Her book implicitly suggested that a truer account of women's bodies was possible, if one could only cut through the morass of misogynistic culture surrounding them. Some feminist scholars have taken this project a step further, asking whether the body exists outside history as a natural object or whether it is fully part of history, necessarily shaped by it.[29] In her chapter, Nelly Oudshoorn concisely sketches the history of recent feminist analyses of the body and, highlighting the example of the as yet nonexistent male contraceptive, makes a strong case for the potential of such analyses to inform the process of historical and political change. She rejects, that is, the contention that deconstructing the meanings of apparently natural objects, like bodies, or of scientific technologies, like male contraceptives, leads feminism away from politics and from the "real" world of bodies and persons. Oudshoorn shows how feminists in the 1970s, laboring in the shadow of the maxim "biology is destiny" and thus wary of engaging the issue of women's nature on scientific grounds, poured their energies into fashioning social critiques of prevailing conceptions of women's nature. Introducing the concept of gender into their critical armamentarium in order to distinguish nature from nurture, an enormously productive move, they left the body untouched, unwittingly characterizing it as natural, beyond history. In the 1980s, feminist scholars began to subject this natural body to scrutiny, proposing that it and its truths were not discovered but, rather, constructed by science. The body, Oudshoorn argues, then entered the feminist research agenda.

But even as feminists submitted the female body to vigorous investigation, they have left the male body largely untouched.[30] Oudshoorn's analysis of the ways in which a naturalized male body has figured in professional and popular discussions of a yet undeveloped male contraceptive suggests there might be benefits to feminism in studying the male body. Time and again, the process of developing a male contraceptive has been short-circuited by appeals to men's essential nature— the supposed complexity of their reproductive systems, their sensitivity to and consequent inability to tolerate pain, their unreliability, and their vulnerability to medical tinkering. Whether or not these characteristics correspond to popular notions of male nature, they have been used successfully to argue for the impossibility of male contraception. As Oudshoorn points out, fifty years of experimentation have yielded thirteen new forms of female contraception, but four hundred years have brought not a single new male method to supplement the condom.

Oudshoorn proposes that the analytical tools feminists have deployed around the female body might be brought to bear on the male body, and that this work of analysis might create new ways of imagining that body, of destabilizing the naturalness and fixity attached to it that make it seem impervious to medical intervention in the service of contraception—intervention to which the female body has been routinely subjected. Oudshoorn calls on feminists to tell new tales, arguing there are plenty of voices out there from which such might be constructed, coming from sources as diverse as the fiction writer on the one hand to the World Health Organization on the other. New tales will bring new meanings, new ways of construing the relations among gendered bodies and medical technologies, and that process, she argues, is ultimately a political one that is both sparked by feminism and will redound to its benefit.

Emily Martin approaches the gendered body from a different perspective, charting the relatively recent rise of a new type of person: the unstable and irrational, yet culturally valued, manic. She shows that whereas formerly the stable, rational, and bounded self was the ideal in Western societies, now, with the irrationalities of the late capitalist system more manifestly on display, flexibility and changeability in the realm of the personality are not only tolerated but even encouraged. Behaviors that used to be condemned as feminine, she argues, are now available to men, even those who ply that most masculine of fields, the market. As new possibilities for personhood emerge, venerable dualities—mind/body, rational/irrational, reason/passion—are called into question, and the strong association of the first of each couplet with masculinity and the second with femininity is destabilized. When stocks can be characterized as "moody," capital as "manic," and Wall Street as "mentally ill," it is clear that this process is well under way and that the predominance of reason is no longer so assured as it once was.

Martin examines the psychiatric category of mania, showing that as it is subjected to conflicting cultural and scientific forces it is being rendered at once less and more concrete and "real." Mania as a way of being in the world, a variation on normal emotionality rather than a discretely bounded illness, can be glimpsed in a range of persons and settings, from politicians to stand-up comics. Widely and variously instantiated, the category loses its former concreteness and boundedness. At the same time, as researchers attempt to locate the sources of manic predispositions in the blood and in the genes, it gains in facticity, appearing less a cultural construct than a timeless entity.

Martin suggests that feminism has been central to the analysis of processes such as the one she chronicles. Feminist social scientists have trained their eyes on all manner of gender and racial hierarchies, exposing both their nonnaturalness and their congruence with various regimes of power. Martin's feminist analysis in this instance is part of a larger project that would subject taken-for-granted inequalities and dichotomies to a range of discipline-specific analytical tools and that would chart the workings of power at the level of the ordinary and the day-to-day. There is a long tradition of feminist critiques of psychiatric knowledge as gendered and psychiatry as misogynistic.[31] Martin's focus here is somewhat different, less on the way that classificatory systems constrain and categorize complex individuality than on the way in which new possibilities for being in the world undermine long-established ways of classifying persons. Feminism, in her chapter, appears as a tool of analysis. One might also ask additionally whether feminism had any role in breaking down the strict dichotomies between male and female norms of personhood that she finds under way, whether, that is, feminism might be a player alongside capitalism in the historical story she tells.

Like Oudshoorn and Martin, Evelynn Hammonds argues for a feminist analysis of the cultural politics of the body that will become part of the history it analyzes. Examining how feminists positioned themselves and the category of "women" when taking on what she argues has been the highly gendered HIV/AIDS epidemic, Hammonds sketches a complex portrait of engagement and activism on the one hand and invisibility and fragmentation on the other. She shows that situating women in representations of an epidemic that was from the start linked to gay men has been a struggle. That women entered official accounts of the epidemic first as prostitutes, as women who put men at risk (not as women who were themselves at risk), was but one telling manifestation of women's larger marginality to the discussion of AIDS, of how infected or affected women until recently were characterized primarily in terms of their relationship to men. Hammonds highlights the marginality of women at a number of sites in the debate, from minority women living in communities which distanced themselves from the homosexuality and drug abuse with which the disease was associated to women working in AIDS organizations that put gay men's concerns front and center—a marginality that was in this case paradoxical in that such organizations drew lessons from an earlier generation of feminist activists who had insisted that personal politics mattered. In the popular media, infected women were categorized as either innocent or guilty, and it was thus impossible to represent them stably.

Questions of representation are not, in Hammonds's hands, divorced from politics. Rather, she argues, they are intimately related, and she further suggests that even as those speaking on behalf of infected women have bridged some of their differences, new omissions and silences have emerged. The racial and class dimensions of the epidemic, especially concerning women, remain more occluded than one might have expected in light of the state of feminist theorizing. Hammonds enjoins us to move women from the margins to the center of the AIDS discussion and suggests that in doing so we will not only produce new analyses but also save lives.

The chapters in this section highlight some of the ways in which medical practice continues to be deeply gendered, despite twenty-five years of feminist theorizing and health activism. Contraception is still considered a women's issue, HIV/AIDS a men's issue. As Oudshoorn and Hammonds show, the effects of this gendering are detrimental to both women and men. Feminist analyses of medical practice have made a difference, as pointed out above; in clinical research, for example, women are no longer routinely considered aberrant compared to the male norm. But, as Martin's chapter reminds us, the workings of gender in medical—including psychiatric—practice are complex and not always readily visible. It may be that the sorts of changes in modal personhood that she charts, with formerly devalued "feminine" traits such as flexibility and adaptability being newly valued, will prove as critical to the project of degendering medicine and medical practice as feminist analyses of such changes.

Feminism has altered the course of science, technology, and medicine in the twentieth century. This "feminism," however, as Ruth Schwartz Cowan has pointed out, is not "univocal." The women's movements of the 1960s took issue not only with men and dominant forms of masculinity but with earlier feminist activists who had championed women's activities as consumers, household managers, and reformers. As Scott Gilbert and Karen Rader have emphasized in their paper, "a feminist scientific agenda of one age might be the reactionary agenda of a different age." [32]

Even today, feminists active in science, technology, and medicine have employed diverse points of view, modes of analysis, and even varying visions of equality between the sexes. Feminism, in our vision, is not rigid and static but flexible and innovative. Neither is it only for women, a position that primatologist Shirley Strum has called "science with pink ribbons." [33] Consequently, in this volume we have cultivated open-ended discussions of many different forms of feminist analysis as they have contributed to mainstream science, technology, and medicine in the twentieth century. It has been our goal to show in some detail how specific modes of feminist analysis have contributed to particular scientific disciplines, aspects of technology, and areas of medicine. Taken together, these examples provide an important basis for contemplating not only the past contributions of feminism to science, technology, and medicine but also its future.

NOTES

1. The feminist scholarship on science, technology, and medicine is too extensive to reference in a note, but several recent collections provide a current perspective on the literature: Evelyn Fox Keller and Helen E. Longino, eds., *Feminism and Science* (Oxford: Oxford University Press, 1996); Sally Gregory Kohlstedt and Helen E. Longino, eds., *Women, Gender, and Science: New Directions*, special issue of *Osiris* 12 (1997); Nina E. Lerman, Arwen Palmer Mohun, and Ruth Oldenziel, guest eds., *Gender Analysis and the History of Technology*, special issue of *Technology and Culture* 38 (1997): 1–231; Annie Canel and Karin Zachmann, guest eds., *Gaining Access, Crossing Boundaries: Women in Engineering in a Comparative Perspective*, special issue of *History and Technology* 14 (1997): 1–157; Rima D. Apple, *Women, Health, and Medicine in America: A Historical Handbook* (New York: Garland, 1990); Judith Walzer Leavitt, ed., *Women and Health in America: Historical Readings* (Madison: University of Wisconsin Press, 1999).

2. As evidence of the tendency to group science, technology, and medicine together, see Paul T. Durbin, ed., *A Guide to the Culture of Science, Technology, and Medicine* (New York: Free Press, 1980); and the ongoing H-NET List on the History of Science, Medicine, and Technology.

3. Margaret W. Rossiter, *Women Scientists in America: Struggles and Strategies to 1940* (Baltimore, MD: Johns Hopkins University Press, 1982); National Science Foundation, *Women, Minorities, and Persons with Disabilities in Science and Engineering: 1998* (Arlington, VA, 1999); Londa Schiebinger, *Has Feminism Changed Science?* (Cambridge, MA: Harvard University Press, 1999).

4. Schiebinger, *Has Feminism Changed Science?* For one cardiologist's story, see Gina Kolata, "The Double Life of Dr. Swain: Work and More Work" and "Twin Sees Herself Stymied," *New York Times*, Sept. 27, 1994, sec. C, p. 1, col. 1, continued on p. 10, col. 1.

5. Rossiter, *Women Scientists in America: Struggles and Strategies to 1940*; Londa Schiebinger, *The Mind Has No Sex? Women in the Origins of Modern Science* (Cambridge, MA: Harvard University Press, 1989); Margaret W. Rossiter, *Women Scientists in America: Before Affirmative Action, 1940–1972* (Baltimore, MD: Johns Hopkins University Press, 1995). For women physicians in the twentieth century, see Regina Markell Morantz-Sanchez, *Sympathy and Science: Women Physicians in American Medicine* (New York: Oxford University Press, 1985), chaps. 10–12. On women engineers, see chap. 8, this volume.

6. Lesley Primmer, "Women's Health Research: Congressional Action and Legislative Gains: 1990–1994," in *Women's Health Research: A Medical and Policy Primer*, ed. Florence P. Haseltine and Beverly Greenberg Jacobson (Washington, DC: Health Press, 1997), 301–30, esp. 302. See also Elizabeth Fee and Nancy Krieger, eds., *Women's Health, Politics, and Power: Essays on Sex/Gender, Medicine, and Public Health* (Amityville, NY: Baywood Publishing, 1994); Sheryl Burt Ruzek, Virginia L. Olesen, and Adele E. Clarke, eds., *Women's Health: Complexities and Differences* (Columbus: Ohio State University Press, 1997).

7. Mary Fainsod Katzenstein, *Faithful and Fearless: Moving Feminist Protest inside the Church and Military* (Princeton, NJ: Princeton University Press, 1998).

8. Linda Marie Fedigan, "Is Primatology a Feminist Science?" in *Women in Human Evolution*, ed. Lori D. Hager (New York: Routledge, 1997), 56–75. For

more on feminism's complex effects on primatology, see Donna Haraway, *Primate Visions: Gender, Race, and Nature in the World of Modern Science* (New York: Routledge, 1989).

9. Schiebinger, conclusion to *Has Feminism Changed Science?*

10. As Wylie points out, a 1997 poll found similar attitudes among the general public: although ever more people than ten years ago embraced equality for women, fewer chose to call themselves feminist (CBS News poll, Sept. 18–20, 1997).

11. See Emily Martin, "The Egg and the Sperm: How Science Has Constructed a Romance Based on Stereotypical Male-Female Roles," *Signs: Journal of Women in Culture and Society* 16 (1991): 485–501; also Biology and Gender Study Group, "The Importance of Feminist Critique for Contemporary Cell Biology," *Hypatia* 3 (1988): 172–87; and Bonnie B. Spanier, *Im/partial Science: Gender Ideology in Molecular Biology* (Bloomington: Indiana University Press, 1995).

12. See Sharon Traweek, *Beamtimes and Lifetimes: The World of High Energy Physicists* (Cambridge, MA: Harvard University Press, 1988); Karen Barad, "Agential Realism: Feminist Interventions in Understanding Scientific Practices," in *The Science Studies Reader*, ed. Mario Biagioli (New York: Routledge, 1999), 1–11; and Amy Bug, "Gender and Physical Science: A Hard Look at a Hard Science," in *Women Succeeding in the Sciences: Theories and Practices across the Disciplines*, ed. Jody Bart (West Lafayette, IN: Purdue University Press, 2000), 221–44.

13. Paul R. Gross and Norman Levitt, *Higher Superstition: The Academic Left and Its Quarrels with Science* (Baltimore, MD: Johns Hopkins University Press, 1994).

14. For current developments in feminist studies of physics, see Schiebinger, *Has Feminism Changed Science?* chap. 9.

15. For an excellent historiographical overview, see Nina E. Lerman, Arwen Palmer Mohun, and Ruth Oldenziel, "The Shoulders We Stand on and the View from Here: Historiography and Directions for Research," *Technology and Culture* 38 (1997): 9–30.

16. For a more recent and quite comprehensive work along these lines, see Autumn Stanley, *Mothers and Daughters of Invention: Notes for a Revised History of Technology* (New Brunswick, NJ: Rutgers University Press, 1995). Judith A. McGaw reviews this and other accounts of women inventors in "Inventors and Other Great Women: Toward a Feminist History of Technological Luminaries," *Technology and Culture* 38 (1997): 214–31.

17. Ruth Schwartz Cowan, "From Virginia Dare to Virginia Slims: Women and Technology in American Life," *Technology and Culture* 20 (1979): 51–63.

18. Judith A. McGaw, "No Passive Victims, No Separate Spheres: A Feminist Perspective on Technology's History," in *In Context: History and the History of Technology*, ed. Stephen H. Cutcliffe and Robert C. Post (Bethlehem, PA: Lehigh University Press, 1989), 172–91.

19. Carroll W. Pursell, "The Long Summer of Boy Engineering," in *Possible Dreams: Enthusiasm for Technology in America*, ed. John L. Wright (Dearborn, MI: Henry Ford Museum and Greenfield Village, 1992), 34–43; Carroll Pursell, "The Construction of Masculinity and Technology," *Polhem* 11 (1993): 206–19.

20. Ruth Schwartz Cowan, "The Consumption Junction: A Proposal for Research Strategies in the Sociology of Technology," in *The Social Construction of Technological Systems: New Directions in the Sociology and History of Technology*, ed. Wiebe E. Bijker, Thomas P. Hughes, and Trevor J. Pinch (Cambridge, MA: MIT Press, 1987), 261–80.

21. See Ronald R. Kline, "Agents of Modernity: Home Economists and Rural Electrification, 1925–1950," in *Rethinking Home Economics: Women and the History of a Profession*, ed. Sarah Stage and Virginia B. Vincenti (Ithaca, NY:

Cornell University Press), 237–52, esp. 239. Oldenziel also draws on several other essays from this volume in her chapter.

22. Mahoney refers here to Ulrike Erb, "Exploring the Excluded: A Feminist Approach to Opening New Perspectives in Computer Science," in *Women, Work, and Computerization: Spinning a Web from Past to Future,* Proceedings of the Sixth International IFIP Conference, Bonn, Germany, May 24–27, 1997, ed. A. Frances Grundy et al. (Berlin and New York: Springer Verlag, 1997), 201–7.

23. See Faye D. Ginsburg and Rayna Rapp, eds., *Conceiving the New World Order: The Global Politics of Reproduction* (Berkeley and Los Angeles: University of California Press, 1995).

24. This volume, p. 135.

25. Barbara Ehrenreich and Deirdre English, *For Her Own Good: 150 Years of the Experts' Advice to Women* (Garden City, NY: Doubleday, Anchor Books, 1978).

26. Boston Women's Health Book Collective, *Our Bodies, Ourselves: A Book by and for Women* (New York: Simon and Schuster, 1973).

27. Carroll Smith-Rosenberg, "The Hysterical Woman: Sex Roles and Role Conflict in Nineteenth-Century America," *Social Research* 39 (1972): 652–78.

28. Emily Martin, *The Woman in the Body: A Cultural Analysis of Reproduction* (Boston: Beacon Press, 1987).

29. See, for example, Barbara Duden, *The Woman beneath the Skin: A Doctor's Patients in Eighteenth-Century Germany,* trans. Thomas Dunlap (Cambridge, MA: Harvard University Press, 1991).

30. For a recent overview, see Susan Bordo, *The Male Body: A New Look at Men in Public and in Private* (New York: Farrar, Straus, and Giroux, 1999).

31. Notably, Phyllis Chesler, *Women and Madness* (Garden City, NY: Doubleday, 1972); and Elaine Showalter, *The Female Malady: Women, Madness, and English Culture, 1830–1980* (New York: Pantheon, 1985).

32. This volume, p. 76.

33. This volume, p. 64.

Science

TWO

Doing Social Science as a Feminist: The Engendering of Archeology

ALISON WYLIE

The question I address in this paper is, What difference has feminism made to archeological research? Archeology is an interesting field to consider in this connection for several reasons. For one, archeology is typically characterized, both by the popular media and internally, as a distinctively masculine enterprise; certainly it is not one of the social or life sciences that have been identified (for better or worse) as "feminized" or prospectively "feminist" fields. More to the point, archeology has shown little evidence of feminist influence until quite recently. Despite its close affiliation with other subfields of anthropology in North America and with neighboring fields in which feminist research has flourished since the early 1970s (e.g., history and paleontology), a comparable body of work has appeared in archeology only in the last decade.

Perhaps because these developments are relatively late, their feminist affiliations are somewhat tenuous; the emerging subfield is typically referred to as "the archeology of gender" or "gender archeology," and a number of those currently active in the area disavow any explicitly feminist commitments. This state of affairs has been cause for concern for those who approach their work as feminists and object that the failure systematically to engage feminist issues and resources limits the potential of research in "the gender genre" to make significant contributions to archeology as a whole.[1] So the question of what contributions feminism has made to archeology is somewhat vexed. I begin by describing the recent and rapid development of archeological research on women and gender. I will then consider what it means to characterize a research program as feminist and assess the contributions of the archeology of gender in these terms.

THE ARCHEOLOGY OF GENDER

Although there are a number of important antecedents to which current work in the archeology of gender can be traced, the first widely influential argument for its potential appeared in 1984.[2] In this essay, Margaret W. Conkey and Janet D. Spector argued that archeology both needs and can sustain a systematic program of feminist research on questions about women and gender. Although such questions had never been central to the research agenda of the discipline, archeologists do routinely make assumptions about sexual divisions of labor, the differentiation of gendered roles (e.g., within households and in larger social institutions and hierarchies), the significance of what they take to be gender symbolism (e.g., in rock art, ceramic decoration, site layout, and architecture), the gender associations of particular classes of tools and related activities, and many other gendered dimensions of the cultures they study. At the very least, Conkey and Spector insisted, archeologists who rely on these assumptions should be prepared to subject them to systematic empirical and conceptual scrutiny; often they underpin established conventions of archeological description, classification, and interpretation that are simply taken for granted.

In addition, Conkey and Spector urged that archeologists make questions about the gendered dimensions of historic and prehistoric cultures a central part of the research agenda of the field. At the time, in the early 1980s, there was virtually no work in archeology they could cite to illustrate the potential of such lines of inquiry; they drew inspiration from traditions of feminist research that were well established in neighboring fields. Feminist sociocultural anthropology, they observed, had already developed through several stages of critique, remediation, autocritique, and ever deepening reassessment of fundamental categories of analysis; it offered substantial resources, empirical and theoretical, for recasting archeological questions and reinterpreting archeological data. They also noted the constructive influence of feminist thinking on "reconstructions of earliest hominid life,"[3] unseating "man the hunter" models of human evolution and opening up a number of hitherto unexplored interpretive possibilities.[4]

Conkey and Spector's article generated intense discussion as soon as it appeared, but it was another seven years before the first major publications began to appear that took up the challenge they had posed. One of the earliest of these, *Engendering Archaeology: Women and Prehistory*,[5] resulted from a small working conference convened by Joan M. Gero and Margaret W. Conkey in April 1988. For most of the archeological participants this was the first time they had considered questions about women or gender. The following year the student organizers of an annual thematic conference at the University of Calgary chose, as their theme for the 1989 Chacmool conference, "The Archaeology of Gender." To everyone's surprise, the open call for papers for this meeting drew over 100 contributions on a wide range of topics, a substantially larger response than had been realized for any previous Chacmool conference. The proceedings of this conference were published two years later, in the same year as *Engendering Archaeology*.[6] The only previous conferences on the archeology of gender had been a Norwegian meeting held in 1979,[7] several sessions on women and gender organized through the 1980s for the annual meetings of the Theoretical Archaeology Group in the United Kingdom,[8] and annual colloquia at the meetings of the Society for Historical Archaeology beginning in 1988.

In the subsequent decade a rapidly expanding number of conference symposia, workshops, and autonomous, special-topic meetings of various scales were organized in Australia, the United Kingdom, the United States, and Europe, many of which have produced edited volumes, special issues of established journals, and published proceedings. For example, Australian archeologists have organized "Women in Archeology" conferences every couple of years since 1990,[9] and at least two conferences on gender have taken place in Europe since 1998 (one at the Bellagio Center and the other in Germany). Beginning in 1991, Appalachian State University in North Carolina has supported a series of spring meetings on the archeology of gender and women in archeology; the proceedings of the first and selected papers from the second are now in print.[10] The School of American Research in Santa Fe recently sponsored an Advanced Seminar entitled "Doing Archaeology as a Feminist,"[11] and a number of special sessions on gender topics have been organized for the annual meetings of national archeology societies, several of which have resulted in publications: the American Anthropological Association,[12] the Society for American Archaeology,[13] and the Society for Historical Archaeology.[14]

It is important to recognize that the emergence of widespread interest in the archeology of gender has been connected in various ways to a parallel and, in most areas, antecedent interest in questions about the roles, status, and contributions of women in archeology. Beginning in the early 1980s a burgeoning literature has reported the results of a number of fine-grained, empirical studies that focus on gender dimensions of the demography, training, employment patterns, and funding of archeologists.[15] The earliest of these was a summary of survey results published by the *Anthropology Newsletter* in 1980, al-

though many of its central points were anticipated by a sharply sa-
tirical discussion of the experience of women in various subfields of
anthropology that appeared in the *Anthropology Newsletter* almost a
decade earlier.[16] Recent studies document not only differential support,
training, and advancement for women in archeology but also entrenched patterns of gender segregation that circumscribe the areas in
which women typically work. Although this equity research tends to
focus exclusively on the status and experience of women—indeed,
women are one of the few traditionally excluded groups to gain sufficient levels of representation within archeology to develop such critiques on their own behalf—it is important to recognize its affiliations
with a much more expansive literature on the "sociopolitics" of archeology that has been concerned principally with the class, race, and
nationalist dimensions of archeological institutions and practice.[17]

The literature that comprises the archeology of gender (including
both equity studies and archeological research on women and gender)
has grown exponentially since 1992, generating a number of extensive
bibliographies,[18] several monographs,[19] and, most recently, an ambitious thematic overview[20] and a reader.[21] To date, "over 500 conference papers authored by over 400 individuals have been presented on
gender since 1988, and over 10 conferences devoted to archaeology
and gender have been held since 1987."[22]

FEMINIST ARCHEOLOGY, EQUITY STUDIES, AND
THE ARCHEOLOGY OF GENDER

It is a striking feature of this growing literature on the archeology of
gender that relatively few contributors explicitly identify themselves
or their research as feminist, and many make little use of the resources
of feminist scholarship developed in neighboring fields. In one of the
first published analyses of this literature, a review of the abstracts for
the 1989 Chacmool conference, Marsha P. Hanen and Jane H. Kelley
describe the breadth and diversity of archeological interests represented at this first open conference but note a dearth of references to
feminist literature, authors, influences, or ideas.[23] This observation was
confirmed by the results of a survey I undertook of participants in this
conference.[24] A substantial majority of those who contributed papers
were women,[25] and three quarters reported a preexisting interest in
questions about gender. The survey respondents also made it clear,
however, that an avowed interest in such questions did not necessarily reflect an underlying commitment to feminism. Nearly half the
women, and more of the men, said explicitly that they did not identify

themselves as feminists, and many of those who embraced the label recorded reservations about what it means.[26] By contrast, when I interviewed those who had played a role as "catalysts"—organizers of the Chacmool or other early conferences, editors of the first publications in the area—I discovered that most had been independently politicized as feminists and had then brought an explicitly feminist angle of vision to bear on the programs of research in which they were engaged as archeologists.

When I undertook this survey of Chacmool participants I assumed that they must have been motivated to attend the conference for the same reasons as had inspired the "catalysts" to undertake their various organizational and editorial projects. That is, I assumed that they were feminists working in archeology who saw the Chacmool conference as an opportunity to bring together, for the first time, their political commitments and their professional, archeological interests; what else would have attracted such a widely dispersed, predominantly female group of archeologists to a conference entitled "The Archeology of Gender" when virtually nothing on the topic had appeared in print? Given the results of the survey, however, I concluded that the interests motivating many participants in the Chacmool conference might best be described as a matter of sensitivity to gender issues, no doubt in some sense a gendered standpoint but not an explicitly feminist standpoint.[27]

Altogether 60 percent of the women who responded to my survey were early to midcareer professionals, having entered the field in the late 1970s and early 1980s, when the representation of women in North American archeology doubled and the first stirrings of interest in the equity issues to do with the status of women in the field were evident. The very presence of these women in archeology—the gender composition of their professional cohort—posed a challenge to conventional assumptions about gender roles that had long structured the professional lives and workplaces of archeologists. However insulated archeologists might be from direct involvement in feminist activism, the second-wave women's movement had disrupted settled assumptions about gender roles and relations in the society at large, and by the late 1970s and 1980s, this was transforming the archeological workplace. The 1989 Chacmool call for papers seems to have resonated with a growing awareness of the contested and contestable nature of gender roles, considered both as a feature of daily life and as a possible topic for investigation in archeology. Indeed, several respondents to my survey reported that it was through the Chacmool conference, and the interest it fostered in scholarly, archeological questions about gender, that they came to see the relevance of feminist activism to their lives.

For the most part, however, feminist commitments continue to be muted in archeology. In a recent review of the archeology of gender, Conkey and Gero remark on the conservatism of the work produced in the last decade, the period in which the archeology of gender has taken shape and achieved prominence in the field as a whole. They describe a tendency to focus on questions about women and gender that are minimally disruptive of established research programs and practices. They observe that the result is "rather thin gruel";[28] much more innovative work would be possible if researchers working in the "gender genre" made full use of the resources (theoretical and empirical) developed by established traditions of feminist scholarship in related fields. Indeed, Conkey and Gero suggest that, even considered in its own terms, the intellectual integrity of "gender research" has been compromised by ignoring these resources.[29]

ON FEMINIST PRACTICE AND PERSPECTIVES

To assess the contributions that *feminism* has made to archeology, then, it is necessary to consider in general terms what constitutes a feminist approach or orientation to research. If one follows Shulamit Reinharz, for example, and focuses on work that is identified by its authors as feminist,[30] much of the recent archeological work on gender would remain unaccounted. Given the foregoing analysis, however, I would argue that feminist commitments and insights have played a critical, if often indirect, role in getting the archeology of gender off the ground in all its forms.

It is perilous, indeed, to attempt to characterize feminism in terms of any very closely defined core of beliefs or values, but it does seem plausible to claim that, as Helen E. Longino suggests, feminist scholars share a broad commitment to a "bottom line requirement": a commitment to "prevent gender from being disappeared."[31] More specifically, feminists generally treat gender as a crucial dimension along which our lives are organized, although it is not necessarily primary and it never operates in isolation from a wide range of other structuring principles. Feminists also typically assume that, as contingent as they are in form and significance, gender roles and relations are hierarchical; those persons categorized as women under existing sex/gender conventions tend to be, in varying ways and to different degrees, systematically disadvantaged *as women*. Finally, insofar as sex/gender systems entrench systemic inequities—insofar as they delimit quite different opportunities, privileges, resources, and rewards for those categorized as women—feminists regard them as unjust; as feminists, they are committed to changing these inequities.[32]

Given these empirical and normative presuppositions, feminists have been interested in science not only as critics who have played a prominent role in documenting myriad ways in which the sciences embody and legitimate gender inequalities but also as practitioners and activists committed to the advancement of science and to the creative use of scientific knowledge to improve our collective lives. In this latter capacity a great many feminists (in particular, feminist social scientists) regard the tools of scientific inquiry as a crucial resource for establishing an empirically well grounded, explanatorily powerful understanding of exactly what forms sex/gender systems take, how they operate, and what (differential) effects they have on our everyday lives. Indeed, the point is frequently made by feminists that, to be effective activists, it is necessary to understand the conditions that disadvantage women with as much empirical accuracy and explanatory power as possible. Far from compromising the integrity of scientific practice, the commitments feminists bring to their research practice often raise the epistemic stakes in ways that improve the sciences they engage.

Although relatively few archeologists identify explicitly as feminists, I would argue that much of the research done under the rubric of the archeology of gender embodies at least a minimal feminist commitment to Longino's "bottom line maxim"; even if this research is not internally identified as feminist, it is animated by a concern to ensure that questions about women and gender are not "disappeared." And while much remains to be done, the commitment to take women and gender seriously has resulted in contributions to archeology that are changing its practice, its research agenda, and its understanding of the cultural past. Let me describe these contributions briefly and return, in my conclusion, to the question of what more explicitly feminist approaches might have to offer.

THE FRUITS OF NOT "DISAPPEARING" GENDER

Often feminist research begins with, and is best known for, its incisive critiques of sexism and androcentrism in extant bodies of knowledge. This provokes the question, But what *constructive* contributions have feminists made to the research fields they engage? I want to resist the assumption, implicit in this question, that critical analyses are a clearly distinct genre and do not represent contributions in their own right—that, as deconstructive of entrenched assumptions, they are primarily

destructive (if that is the intended contrast). In practice, critical interventions often depend upon the systematic restudy of a domain or
problem or body of data, and frequently they are the catalyst for reframing questions that open up fruitful new lines of inquiry. But even
when feminist critique is primarily negative, it must be counted a significant contribution to show where settled assumptions are problematic and to clearly identify the explanatory and empirical inadequacies
of widely accepted knowledge claims; in this, feminist practice embodies just the kind of rigorous critical engagement that distinguishes the
best of scientific practice. In the brief catalogue of putatively feminist
contributions to archeology that follows, I first consider some forms
of critical analysis that reflect a commitment to counter the disap-
pearing of gender and then describe a selection of questions, research
strategies, background assumptions, and theoretical insights to which
these have given rise. Although differences of emphasis are clear
enough, the examples themselves subvert any hard and fast categorization; some serve to illustrate several different kinds of contribution,
and all of them illustrate the interdependence between critical analysis
and constructive insight.[33]

Countering Erasure

One standard form of feminist critique in the social sciences is the
documentation of systematic erasure, where women and gender are
simply left out of the account or are assumed to have been taken into
account when male-associated experience, interests, and activities are
generalized to the subject domain as a whole. There has been a great
deal of work along these lines in archeology in the last decade. In one
particularly striking and explicitly feminist example, Gero calls into
question the research agenda of a whole archeological subfield.[34] She
argues that the defining problematic of Paleo-Indian research is an
artifact of a marked preoccupation, among predominantly male practitioners, with stereotypically male activities: large-scale mammoth-
and bison-hunting practices and associated kill sites and hunting-tool
assemblages. This is a preoccupation that marginalizes lines of inquiry—forms of evidence and analysis—that are central to traditions
of research on subsistence practices in many other areas. There is evidence that Paleo-Indians foraged a wide range of plant materials, as
well as hunting Pleistocene megafauna, but it comes from studies conducted largely by women on expedient flake tools associated with the
gathering and food-processing activities of women. In a study of cita-
tion patterns, Gero finds that this work remains outside the mainstream in the field of lithic studies. The questions central to Paleo-

Indian studies are, consequently, questions about what happened to the mammoth hunters when these Pleistocene fauna disappeared: was the mammoth-hunting population replaced by foraging groups who relied primarily on small game and plant resources, or did they effect a miraculous transformation as their subsistence base changed? Gero argues that these questions can arise, in these terms, only if you ignore the evidence—from research by women on tools associated with women's activities—that Paleo-Indians depended on a much more diversified set of subsistence strategies than acknowledged by standard "man the (mammoth/bison) hunter" models. In short, the preoccupation with male hunting activities seems to have obscured a much more diversified and flexible subsistence pattern than has typically been attributed to the earliest inhabitants of the Americas. In a recent analysis, Gero extends these insights to the primary contexts of archeological data recovery, showing how the gendered organization of labor in the field can structure the very form and content of the archeological "facts" established by excavation.[35]

In a similar vein, critical reanalyses of dominant theories about the Inka and Aztec states make it clear that our understanding of state formation processes is seriously flawed if these processes are identified exclusively with structural dynamics operating in the public, political sphere associated predominantly with male elites. Christine A. Hastorf provides compelling evidence (based on analyses of skeletal material, paleobotanical remains, and a sequence of house floors in the Montaro Valley) that the extension of Inka influence into the highland Andes substantially reshaped the domestic sphere associated with women and gender relations more generally; the household forms encountered at the time of the Spanish conquest cannot be projected back into prehistory as if their form was a given, a social foundation unaffected by the rising and falling fortunes of larger state systems.[36] Elizabeth M. Brumfiel makes the case, not just that the Aztec system of economic and political control changed domestic relations, but that it depended fundamentally on the intensified and restructured exploitation of women's (domestic) labor, given its reliance on a practice of extracting tribute in the form of cloth produced by women. She bases this reassessment on an analysis of changes, over time, in the distribution of artifacts associated with food and cloth production across urban and hinterland sites in central Mexico Valley.[37]

In a similar example drawn from historical archeology, Anne Yentsch argues that the organization of domestic labor cannot be assumed to be external and marginal to the large-scale processes of industrialization that transformed the organization of production in the northeastern United States in the eighteenth and nineteenth centuries.[38] On her account, largely unexamined processes of appropriation of "women's work" played a key role in creating markets for new products and transforming the rural economy in this period. In one case that she examines in some detail, Yentsch delineates previously unrecognized patterns of change in the ceramic ware found in domestic assemblages that testify to the gradual transfer of women's productive activities (specifically, domestic dairy production) from the household to commercial enterprises whenever these proved amenable to industrialization.

Countering Stereotypes

Often problems of gender bias in archeology are not so much a consequence of simple erasure as of dependence on unexamined stereotypes about the capacities and roles proper to women and men. This is evident in all aspects of archeological inquiry, ranging from the recovery, description, and interpretation of data as evidence through to reconstructive models and explanatory theories. Whatever the subject or scale of analysis, the default assumption often seems to be that gender conventions familiar from contemporary (white, middle-class, Western) contexts can be projected onto the cultural past as if they were in some sense "natural," a stable, universal substrate of social, cultural life.

One example of how gender stereotypes can structure archeological inquiry at a fundamental empirical level comes from a reanalysis of skeletal material recovered from archeological contexts in Australia. Denise Donlon notes that, as currently catalogued, existing skeletal collections show an implausible preponderance of male specimens.[39] She canvasses various possible explanations for this—demographic anomalies, differential preservation, curatorial practices—and concludes that the skewed sex ratios reported for Aboriginal skeletal remains probably reflect systematic errors in sex identification due, specifically, to pervasive reliance on measures of "robustness." Given what is known historically and ethnographically about the activities typical of Aboriginal women, she argues, it is more than likely that they would show much the same levels of skeletal robustness as their male counterparts; it is a mistake to assume that sex difference stereotypes about physical dimorphism that reflect our own highly gender-segregated activities can be extended to prehistoric foragers in Australia.

A parallel objection arises in connection with a great many reconstructive models: they tend to cast women in passive and dependent roles, even when there is strong archeological or collateral (e.g., ethnohistoric) evidence to the contrary. This is especially clear in the conventions that structure visual reconstructions of human activities in deep prehistory, in which, until quite recently, females figure primarily in the background, gaze averted, their subordinate status signaled by posture, task, and affiliation, while males are cast in active, dominant roles; Diane Gifford-Gonzalez offers an analysis of these conventions in a discussion entitled "The Drudge-on-the-Hide."[40] The same themes are evident in models of the subsistence practices of prehistoric foragers, particularly those that have informed evolutionary theorizing. Here conventional models of seasonal subsistence rounds, and associated attributions of function to tools and to sites, reflect an assumption that it was the men who ranged widely to hunt while women (and children) were bound to home bases. This flies in the face of counterevidence from ethnographic research (widely discussed since the early 1970s) that women are often as mobile as men, that their foraging activities may account for the majority of the dietary intake of their communities, and that, as a consequence, they routinely play a key decision-making role in determining group composition and movement.[41]

These gender stereotypes also structure ambitious explanatory accounts of major transformations in technology and subsistence. For example, Kenneth E. Sassaman objects that the transition from mobile to sedentary prehistoric societies in North America should not be attributed to changes in the organization of hunting practices associated with men: "considering . . . that the change coincides with the adoption of pottery, technology usually attributed to women, an alternative explanation must be considered."[42] In a similar vein, Patty Jo Watson and Mary C. Kennedy argue that, however much they differ from one another, dominant explanations of the emergence of horticulture in the Eastern Woodlands of North America (approximately 7000–2000 years ago) tend to deny women agency even in roles that are conventionally ascribed to them.[43] Although women are assumed to have had the primary responsibility for collecting wild plants when foraging practices prevailed, and for tending them when horticulture was established, it is men who are cast in the role of catalysts and innovators whenever human agency is invoked to explain the transition from a foraging to a horticultural subsistence base. On one model, male shamans developed domesticates in the course of experimenting with gourds to produce ritual rattles. On other models, human activities in the area of "domestilocalities" unintentionally selected for the characteristics that later came to distinguish cultivated plants; the plants are assumed to have domesticated themselves. Watson and Kennedy say they are "leery" of explanations that remove women from the one domain granted them as soon as an exercise of initiative is envisioned.

A complementary set of critiques suggest that it may be a mistake, not just to assume that particular clusters of attributes are invariably associated with gender categories, but to assume that these categories were always marked, that past divisions of labor, symbolic imagery, and social roles and relations (for example) were structured by the kind of gender segregation familiar from contemporary contexts. Kelley R. McGuire and William R. Hildebrandt raise this possibility in a critique of models of the transition from a millingstone/handstone technology to bedrock mortars in Californian prehistory. They find that accounts of this transition typically assume that women would have had the primary responsibility for milling major plant resources (e.g., acorns), and yet, they note, ethnographic precedent and archeological evidence suggest that "a consistent pattern of relatively undifferentiated gender roles" may have obtained in foraging practices throughout this period. McGuire and Hildebrandt conclude that "if we have learned anything [from] the emerging feminist critique of modern archeology, it is perhaps the danger in viewing gender relationships as static, or at least limited in range."[44]

In a very different archeological context, that of the comparative interpretation of Minoan and Mycenaean art and architecture, Lucia Nixon raises similar objections to the tendency to interpret prehistoric cultures in terms of "our own presentist, polarized" assumptions about gender roles and attributes. It is a mistake, she urges, to assume that the symbolic prominence of women in Minoan imagery is indicative of a matriarchal society that can be sharply contrasted to the presumed patriarchy of Mycenaean society; this "misplaced gender polarity" obscures significant and instructive similarities between these Aegean Bronze Age societies.[45] Rosemary A. Joyce takes a parallel observation as the point of departure for proposing a systematic reinterpretation of the gender imagery of the Classic Maya.[46] When multiple lines of evidence are considered, contradictions emerge which undercut the presumption that the highly stylized representation of gender dichotomies on monuments represents the actual roles and status of (elite) women in Mayan society. Joyce recommends a "performative" conception of gender symbolism which recognizes that the elite women may have played more diverse and powerful roles than conventional interpretation allows.[47]

Alternative Accounts, Theories, and Strategies of Inquiry

As the foregoing examples suggest, gender critiques in archeology typically embody a commitment to improve on existing practice; they bring into view the implications of taking women and gender into account and thinking beyond ethnocentric (sexist and androcentric) stereotypes. Although there is little explicit debate on the question of what this requires methodologically or theoretically, there does seem to be broad consensus among the proponents of "gender research" that it is a mistake to blame the tools of archeological inquiry for the androcentric or sexist bias of its products; there is no feminist or gender-sensitive methodology that can be counted on to correct or protect against these errors.[48] Most archeologists working in the "gender genre" make creative use of the existing tools of archeological inquiry to establish evidential claims, reconstructive models, and explanatory theories that are gender inclusive.

In many cases questions about women and gender have mobilized a newly insistent interest in documenting activities associated with women that have been assumed to be archeologically inaccessible— that is, repetitive domestic activities involving perishable materials and utilitarian tools that often leave little durable archeological record. For example, there has been a growth of interest in archeological evidence for netting and basketry industries. Sometimes this involves closer, more systematic examination of surviving fragments of such material that have remained scattered and marginal to mainstream archeological analysis,[49] but often it also requires the creative use of indirect methods of analysis. To this end, Marcia-Anne Dobres focuses on the presence of bone awls as evidence of the prehistoric nets and fishing lines they would have been used to make, when nothing of these latter artifacts survives in the archeological record.[50] In a similar vein, Gifford-Gonzalez explores the potential of microanalysis of breakage patterns in bones to establish evidence of domestic (female-associated) activities of food processing rather than primary (male-associated) butchering activities;[51] Cheryl Claassen considers various possibilities for investigating the shellfishing activities of women and children;[52] and Gero urges renewed attention to certain patterns of edgewear in utilitarian food-processing tools that might provide a fuller understanding of women's activities that have largely been ignored.[53]

Another area where new lines of evidence have led to quite striking insights about gender roles and relations is in the analysis of skeletal material. The case that Hastorf makes for reconsidering standard models of Inka state formation (described above) depends in part on a comparison of the results of previous studies in which skeletal material from the Montaro Valley had been sexed and lifetime dietary profiles reconstructed using isotope analysis.[54] Until Hastorf asked whether these dietary profiles differ by sex, it had not been noticed that there is a striking divergence between the dietary intake of males and females during the period when the Inka state was expanding; males begin to show much higher levels of the isotope associated with maize only after the Inka establish a presence in the Andean highlands. Combined with evidence of parallel changes in the intensity with which various food resources were being exploited and processed during this period, this divergence in dietary profiles was crucial in suggesting that the Inka presence had had a significant impact on social relations at the level of the household and local community.

The skeletal component of Hastorf's study represents one strategy of analysis exploited by a growing number of archeologists and physical anthropologists who are interested in exploring osteological evidence of the possible difference along sex lines of routine activities, dietary status, and vulnerability to violence and disease. Useful overviews of these initiatives are provided by Mark N. Cohen and Sharon Bennett and by Gillian R. Bentley; these overviews include, for example, analyses of the differential development of muscle attachments and evidence of stress that result from specific kinds of repetitive movement, evidence that may make it possible to determine whether, in fact, certain tasks were performed predominantly by one sex or by the other.[55]

For the most part the development of these lines of evidence depends not so much on collecting different ranges of data than have been considered archeologically salient in the past, as on innovative approaches to the analysis of existing, often well understood archeological assemblages. As with Hastorf's comparative analysis of the sex: dietary profile data, Brumfiel's thesis about the impact of the Aztec state on domestic labor turns on an analysis that brought to light new patterns of correlation and distribution in well-documented assemblages of spindle whorls and cooking pots.[56] The power of her argument derives in part from the fact that the methods of analysis and the background knowledge she relies on to establish these patterns are uncontroversial for those she engages in debate about larger theoretical issues. Often, however, gender researchers find that, to ask new questions of archeological data, they must exploit somewhat different strategies of analysis and ranges of background knowledge—different auxiliary hypotheses—than have been conventional in their areas of interest. In some cases the necessary resources exist in neighboring fields on which archeologists routinely rely for interpretive principles. For example, the basis for reframing hunting-focused reconstructions of prehistoric foraging practices, and associated "man the hunter" models of human evolution, is a well-established body of ethnographic research which demonstrates (as indicated above) that the "gathering" activities associated with women are often the primary, and most reliable, source of dietary intake for foraging groups.[57] In other cases, however, the relevant background knowledge does not exist, or at least not in a form that can be readily applied to archeological problems. Ethnographic accounts of foraging practices may not include enough detail about the associated material culture to support fine-grained interpretation of the archeological record of prehistoric foragers.

This problem is by no means unique to archeological research on gender. Indeed, one of the distinctive features of contemporary (anthropological) archeology has been its emphasis on innovative programs of research in "ethnoarcheology" and "experimental archeology": materials research, ethnographic and behavioral studies, and replication projects specifically designed to establish the kind of background knowledge about the production, use, and deposition of material culture that archeologists need to make sense of the evidence with which they work.[58] Those engaged in gender research often find it necessary to extend this vigorous tradition of "middle-range research"; they make it a priority to develop the background knowledge necessary to recognize and interpret archeological evidence of gender roles and relations. In one study that takes up this challenge specifically with reference to foraging practices, Hetty J. Brumbach and Robert Jarvenpa document the subsistence practices of the subarctic Dene.[59] They report much more complex social patterns in the organization of hunting activities than had been recognized; in particular, contrary to conventional assumptions, they find that Dene women are routinely involved in certain kinds of hunting parties. This insight has wide implications for the interpretation of archeological sites and assemblages in subarctic environments, and it requires, more locally, a systematic reassessment of conventional accounts of the processes by which historic period Dene shifted their emphasis from "bush"- to "village"-centered subsistence activities.

In one of the earliest proposals for feminist research in archeology, Spector outlined a "task differentiation" framework for behavioral analysis that was designed to ensure that the material correlates of gender-specific activities would be recognized in both ethnohistoric and archeological contexts.[60] Although Spector is now skeptical about its usefulness, something like this approach is evident in Brumbach and Jarvenpa's work and has been widely adopted by ethnoarcheologists intent on developing a systematic basis for interpreting the gendered dimensions of the archeological record.

It is at the level of reframing or developing substantially new reconstructive and explanatory models of the cultural past that gender archeology most clearly requires the resources of sophisticated feminist theory. Conkey and Gero, among others, make an important point when they observe that, to date, "the archeology of gender has been primarily an enterprise of locating women" and adding gender; it has not yet taken the full measure of the implications of insisting that gender not be disappeared.[61] As Sassaman argues in connection with the transition to sedentism, and Brumfiel with respect to state formation processes, it is sometimes necessary to rethink conventional explanatory models quite fundamentally if they are to deal adequately with the insight that sex/gender systems may have taken quite different forms in the past than they do now and may have been an important force in shaping cultural systems.

Sassaman and Brumfiel have both pursued the implications of these insights, exploring more complex (multidimensional and multifactor) explanatory models than have been typical in their areas of interest. In particular, Brumfiel has argued for a thoroughgoing reassessment of the ways in which archeologists conceptualize factional dynamics in complex societies.[62] Others have made a case for shifting the scale of inquiry from a focus on regional cultures to fine-grained analyses of features within sites, reconceptualizing the units of archeological analysis in terms that make individual agency and family- or community-level social relations accessible to archeological analysis. Ruth E. Tringham argues, in this connection, that "household archeology" holds particular potential for feminist projects.[63] The question of what explanatory, theoretical insights are likely to result from this redirection of attention remains open, but Tringham, among others, resists the impulse to replace the master narratives of androcentric theories with new, comparably expansive, feminist or gender-sensitive theories.

CONCLUSION

The archeologists responsible for this growing body of gender research have been very effective in countering the erasure of women and distortions due to gender stereotypes at the level of data description, background assumptions, and reconstructive claims about particular past events or conditions of life in virtually all areas of archeological activity. Although wider implications are often noted, for the most part they remain to be developed. The greatest strength of the archeology of gender—its ubiquity—may be a function of the very conservatism remarked upon by Conkey and Gero.[64]

At the same time, the growing interest in the archeology of gender has had another kind of impact on the field that must be reckoned among its most important contributions. Whether its proponents intend it or not, research in the "gender genre" draws attention to the contingency and defeasibility of settled traditions of practice. Critical analyses throw into relief unrecognized androcentric and sexist assumptions that deeply structure, and compromise, archeological inquiry, while constructive initiatives bring into view a world of investigative possibilities that has been largely ignored. However cautious it is, gender research destabilizes whatever remains of an "innocent" confidence in entrenched conventions that define what counts as a significant problem, a well-formed research program, secure background knowledge, and plausible reconstructive or explanatory models.[65] In this the proponents of an archeology of gender, especially the feminists among them, add to a growing range of voices that have been insisting that archeologists must take responsibility for the normative (political, ethical) commitments, as well as the theoretical assumptions and methodological standards, that structure their practice. They urge that archeologists make reflexive critique an essential component of their research enterprise.

A commitment to reflexivity is evident in many aspects of the work undertaken, in the last decade, by feminists in archeology. Research on the status of women in archeology is perhaps most obvious in the tradition of "critical archeology," as are the critiques of gender stereotypes. In addition, however, feminist practitioners demonstrate an intense interest in rethinking archeological practice on a number of other dimensions that may seem, at first, tangential to the goal of ensuring that gender is not disappeared as a subject of inquiry. They explore strategies for changing the way archeological labor is organized so that it is less hierarchical and more democratic in ways that ensure the effective participation, not only of students and junior colleagues, but of descendant communities, avocational archeologists, and a range of other interested publics. Spector offers an especially compelling account of a research project that was substantially enriched by these initiatives.

Similar themes emerge in many other contexts.[66] For example, feminist archeologists routinely insist that, as feminists, it is important to explore alternatives to the pedagogical practices typical of archeological classrooms and museums.[67] In this spirit, they have taken particular initiative in exploiting nonstandard media and forms of presentation in making their results public. Their commitment here, as elsewhere, is to ensure that archeological data and interpretations will reach a wider range of audiences, and that these audiences will be actively engaged in the process of rethinking what we understand about the cultural past. Spector experiments with intellectual biography and fictionalizing narratives; Tringham with dialogue; and Tringham and Joyce explore the potential of hypertext media to make clear the open-ended nature of archeological practice and its products.[68]

These experiments with alternative ways of doing and presenting archeology are among the most significant, and potentially the most transformative, feminist contributions to the field. They are a matter of doing archeology differently in ways that serve, not just to repair elisions and distortions, but to establish forms of practice that bring a range of perspectives to bear on the archeological enterprise. It is the very fractiousness, the critical self-consciousness, of these feminist-inspired forms of practice that mitigates against the tendency to disappear not only gender but any number of other aspects of the cultural past that have been marginal in archeology to date.

NOTES

1. Margaret W. Conkey and Joan M. Gero refer to this growing body of research as "the gender genre" in their article "Programme to Practice: Gender and Feminism in Archaeology," *Annual Review of Anthropology* 26 (1997): 414.

2. This brief history of gender research in archeology is adapted from Alison Wylie, "The Engendering of Archaeology: Refiguring Feminist Science Studies," *Osiris* 12 (1997): 80–99.

Some earlier publications that address archeological questions about women and gender include an early textbook coauthored by an archeologist (M. Kay Martin and Barbara Voorhies, *Female of the Species* [New York: Columbia University Press, 1975]); a discussion of the role archeology might play in the study of state formation processes that was published in 1977 (Rayna Rapp, "Gender and Class: An Archaeology of Knowledge concerning the Origin of the State," *Dialectical Anthropology* 2 [1977]: 309–16); two archeological papers that appeared in a collection of anthropological and historical studies of Plains Indian women (Alice B. Kehoe, "The Shackles of Tradition," and Janet D. Spector, "Male/Female Task Differentiation among the Hidatsa: Toward the Development of an Archaeological Approach to the Study of Gender," both in *The Hidden Half: Studies of Plains Indian Women,* ed. Patricia Albers and Beatrice Medicine [Washington, DC: University Press of America, 1983], 53–73 and 77–99); Anne Barstow, "The Uses of Archaeology for Women's History: James Mellaart's Work on the Neolithic Goddess at Catal Huyuk," *Feminist Studies* 4, no. 3 (1978): 7–17; and several contributions to a symposium organized for the 1987 annual meetings of the American Anthropological Association that later appeared in the collection *Powers of Observation: Alternative Views in Archeology*, ed. Sarah M. Nelson and Alice B. Kehoe, Archaeological Papers of the American Anthropological Association, no. 2 (Washington, DC: American Anthropological Association, 1990). In 1984 Margaret W. Conkey and Janet D. Spector's essay "Archaeology and the Study of Gender" appeared in *Advances in Archaeological Method and Theory*, vol. 7, ed. Michael B. Schiffer (New York: Academic Press, 1984), 1–38.

3. Conkey and Spector, "Archaeology and the Study of Gender," 6–8.

4. These have been pursued and are described retrospectively by contributors to Lori D. Hager, ed., *Women in Human Evolution* (London: Routledge, 1997).

5. Joan M. Gero and Margaret W. Conkey, eds., *Engendering Archaeology: Women and Prehistory* (Oxford: Basil Blackwell Press, 1991).

6. Dale Walde and Noreen D. Willows, eds., *The Archaeology of Gender: Proceedings of the Twenty-second Annual Chacmool Conference* (Calgary, AB: Archaeological Association of the University of Calgary, 1991).

7. The proceedings were published eight years later: Reidar Bertelsen, Arnvid Lillehammer, and Jenny-Rita Naess, eds., *Were They All Men? An Examination of Sex Roles in Prehistoric Society*, AmS-Varia 17 (Stavanger: Arkeologisk Museum I Stavanger, 1987).

8. See Karen Arnold, Roberta Gilchrist, Pam Graves, and Sarah Taylor, "Women in Archaeology: Special Issue," *Archaeological Reviews from Cambridge* 7, no. 1 (1988): 2–8.

9. Hilary duCros and Laurajane Smith, eds., *Women in Archaeology: A Feminist Critique*, Australian National University Occasional Papers (Canberra: Australian National University, 1994); and Jane Balme and Wendy Beck, eds., *Gendered Archaeology: The Second Australian "Women in Archaeology" Conference* (Canberra: ANH Publications, Australian National University, 1995).

10. Cheryl Claassen, ed., *Exploring Gender through Archaeology: Selected Papers from the 1991 Boone Conference*, Monographs in World Archaeology, no. 11 (Madison, WI: Prehistory Press, 1992); and Cheryl Claassen, ed., *Women in Archaeology* (Philadelphia: University of Pennsylvania Press, 1994).

11. Elizabeth M. Brumfiel, "Doing Archaeology as a Feminist: A Report from the April Collective," paper presented at the Ninety-seventh annual meeting of the American Anthropological Association, Philadelphia, 1998; and Alison Wylie and Margaret W. Conkey, "Doing Archaeology as a Feminist," conference report submitted to the School of American Research, 1998. Both papers are available on the April Collective webpage, "Practicing Archaeology as a Feminist": http://www/qal.berkeley.edu/~aprilfem/SARwelcome.html.

12. Margaret C. Nelson, Sarah M. Nelson, and Alison Wylie, eds., *Equity Issues for Women in Archeology*, Archaeological Papers of the American Anthropological Association, no. 5 (Washington, DC: American Anthropological Association, 1994).

13. Rita P. Wright, ed., *Gender and Archaeology* (Philadelphia: University of Pennsylvania Press, 1996).

14. Donna J. Seifert, ed., *Gender in Historical Archaeology*, special issue of *Historical Archaeology* 25, no. 4 (1991).

15. This account of equity research on the status of women in archeology is developed in more detail in Alison Wylie, "The Trouble with Numbers: Workplace Climate Issues in Archaeology," in Nelson, Nelson, and Wylie, *Equity Issues for Women in Archaeology*, 65–72. For representative examples of equity research, see the other contributions to that volume.

16. Leslie E. Wildeson, "The Status of Women in Archaeology: Results of a Preliminary Survey," *Anthropology Newsletter* 21, no. 5 (1980): 5–8; Anonymous, "The Female Anthropologist's Guide to Academic Pitfalls," *Anthropology*

Newsletter 12, no. 4 (1971): 8–9. Both are reprinted in Nelson, Nelson, and Wylie, *Equity Issues for Women in Archaeology*.

17. For example, Gero published her earliest analysis of gender segregation in the organization of archeological research, "Gender Bias in Archaeology: A Cross-cultural Perspective," in *The Socio-politics of Archaeology*, ed. Joan M. Gero, David M. Lacy, and Michael L. Blakey, Anthropology Research Report no. 23 (Amherst: Department of Anthropology, University of Massachusetts, 1983), a study that predates her first major publications on gender as a subject of archeological research by almost a decade. (See also Joan M. Gero, "Socio-politics and the Woman-at-Home Ideology," *American Antiquity* 50, no. 2 [1985]: 342–50.)

The broader tradition of research on the sociopolitics of archeology is represented by, for example, Kelley and Hanen's analysis of archeological recruiting and training practices that maintain the demographic homogeneity of the discipline on a number of dimensions (Jane H. Kelley and Marsha P. Hanen, *Archaeology and the Methodology of Science* [Albuquerque: University of New Mexico Press, 1988], chap. 4); Patterson's account of ways in which the interests of intranational elites have shaped archeology (Thomas C. Patterson, "The Last Sixty Years: Toward a Social History of Americanist Archaeology in the United States," *American Anthropologist* 88 [1986]: 7–22; and Thomas C. Patterson, "Some Postwar Theoretical Trends in U.S. Archaeology," *Culture* 11, no. 1 [1986]: 43–54) and his analysis of the impact of GI bill educational support on the class structure of the discipline (Thomas C. Patterson, *Toward a Social History of Archaeology in the United States* [Orlando, FL: Harcourt Brace, 1995]); Trigger's wide-ranging study of the alignment of archeology with nationalist agendas of various sorts (Bruce G. Trigger, *A History of Archaeological Thought* [Cambridge: Cambridge University Press, 1989]); and, finally, the range of studies of the institutional dynamics and political economy of archeology that appeared in 1983 in Gero, Lacy, and Blakey, *The Socio-politics of Archaeology*.

18. For example, Elisabeth A. Bacus et al., eds., *A Gendered Past: A Critical Bibliography of Gender in Archaeology*, Technical Report 25 (Ann Arbor: University of Michigan Museum of Anthropology, 1993); and Kelley Hays-Gilpin and David S. Whitely, eds., *Reader in Gender Archaeology* (New York: Routledge, 1998).

19. Janet D. Spector, *What This Awl Means* (Minneapolis: Minnesota Historical Society, 1994); Roberta Gilchrist, *Gender and Material Culture: The Archaeology of Religious Women* (London: Routledge, 1993); and Diana di Z. Wall, *The Archaeology of Gender: Separating the Spheres in Urban America* (New York: Plenum Press, 1994).

20. Sarah M. Nelson, *Gender in Archaeology: Analyzing Power and Prestige* (Walnut Creek, CA: AltaMira Press, 1997).

21. Hays-Gilpin and Whitely, *Reader in Gender Archaeology*.

22. Conkey and Gero, "Gender and Feminism in Archaeology," 414, citing Cheryl Claassen, personal communication.

23. Marsha P. Hanen and Jane H. Kelley, "Gender and Archaeological Knowledge," in *Metaarchaeology: Reflections by Archaeologists and Philosophers*, ed. Lester Embree (Boston: Reidel, 1992), 195–227. Hanen and Kelley find that the topics taken up by Chacmool presenters tended to cluster in or near the historic period but were widely dispersed geographically. Despite this clustering, they conclude that "archaeologists seem to be able to include some kind of gender perspective for whatever geographical area they normally study" (ibid., 206).

24. The results of this survey are reported in Wylie, "The Engendering of Archaeology."

25. Altogether, 72 percent of the 107 participants listed on the program of the 1989 Chacmool conference responded to my inquiries. Of the 107 papers

listed in the program, 80 percent were presented by women; this more than inverts the ratio of women to men in North American archeology, which stood at 36 percent in 1988 when the Chacmool call for papers was circulated. See ibid.

26. This dissociation from feminism is reflected in a number of other responses. Fewer than half the respondents said they had had any previous involvement in women's studies or familiarity with feminist research in other fields, and just half the women and a quarter of the men indicated any involvement in action on women's issues or in "feminist activism"; most described this involvement as limited to "being on a mailing list" or "sending money," usually to women's shelters and reproductive-rights groups. Ibid.

27. In "Gender and Archaeological Knowledge," Hanen and Kelley describe this orientation as a largely untheorized and apolitical "grass-roots" interest in questions about gender relations and categories.

28. Conkey and Gero, "Gender and Feminism in Archaeology," 424, quoting Barbara Bender.

29. Ibid., 424–25.

30. Shulamit Reinharz, *Feminist Methods in Social Research* (Oxford: Oxford University Press, 1992), 4.

31. Helen E. Longino, "In Search of Feminist Epistemology," *Monist* 77, no. 4 (1994): 472–85; and Alison Wylie, "Doing Philosophy as a Feminist: Longino on the Search for a Feminist Epistemology," *Philosophical Topics* 23, no. 2 (1995): 345–58.

32. For a more detailed account of these feminist commitments, see Alison Wylie, "Good Science, Bad Science, or Science as Usual? Feminist Critiques of Science," in Hager, *Women and Human Evolution,* 29–55.

33. Several of the examples I summarize in the sections that follow are analyzed in greater detail in Alison Wylie, "The Constitution of Archaeological Evidence: Gender Politics and Science," in *The Disunity of Science,* ed. Peter Galison and David J. Stump (Stanford, CA: Stanford University Press, 1996), 311–43; and Wylie, "The Engendering of Archaeology."

34. Joan M. Gero, "The Social World of Prehistoric Facts: Gender and Power in Prehistoric Research," in duCros and Smith, *Women in Archaeology,* 31–40.

35. Joan M. Gero, "The Archaeological Practice and Gendered Encounters with Field Data," in Wright, *Gender and Archaeology,* 251–80.

36. Christine A. Hastorf, "Gender, Space, and Food in Prehistory," in Gero and Conkey, *Engendering Archaeology,* 132–59.

37. Elizabeth M. Brumfiel, "Weaving and Cooking: Women's Production in Aztec Mexico," in Gero and Conkey, *Engendering Archaeology,* 224–53.

38. Anne Yentsch, "Engendering Visible and Invisible Ceramic Artifacts, Especially Dairy Vessels," in *Gender in Historical Archaeology,* ed. Donna J. Seifert, special issue of *Historical Archaeology* 25, no. 4 (1991): 132–55.

39. Denise Donlon, "Imbalance in the Sex Ratio in Collections of Australian Aboriginal Skeletal Remains," in duCross and Smith, *Women in Archaeology,* 98–103.

40. Diane Gifford-Gonzalez, "You Can Hide but You Can't Run: Representation of Women's Work in Illustrations of Paleolithic Life," *Visual Anthropology* 9 (1995): 3–21; Diane Gifford-Gonzalez, "The Drudge-on-the-Hide," *Archaeology* 48, no. 2 (1995): 84. For a more expansive analysis, see Stephanie Moser, *Ancestral Images: The Iconography of Human Origins* (Phoenix Mill, U.K.: Sutton Publishing, 1998).

41. See, for example, Nelson's account of such critiques in Nelson, *Gender in Archaeology,* 88–91.

42. Kenneth E. Sassaman, "Lithic Technology and the Hunter-Gatherer Sexual Division of Labor," *North American Archaeologist* 13 (1992): 249.

43. Patty Jo Watson and Mary C. Kennedy, "The Development of Horticul-

ture in the Eastern Woodlands of North America: Women's Role," in Gero and Conkey, *Engendering Archaeology*, 255–75.

44. Kelley R. McGuire and William R. Hildebrandt, "The Possibilities of Women and Men: Gender and the California Millingstone Horizon," *Journal of California and Great Basin Archaeology* 16, no. 1 (1994): 51, 52.

45. Lucia Nixon, "Gender Bias in Archaeology," in *Women in Ancient Societies*, ed. Leonie J. Archer, Susan Fischeler, and Maria Wyke (London: Macmillan, 1994), 9, 8.

46. Rosemary A. Joyce, "The Construction of Gender in Classic Maya Monuments," in Wright, *Gender and Archaeology*, 167–95.

47. Similar criticisms are directed against popular interpretations of Paleolithic and Neolithic "goddess figurines." These figurines were originally identified as evidence of a prehistoric matriarchy by nineteenth-century evolutionary theorists, but the central themes of these early accounts were taken up in the 1970s and 1980s by cultural feminists intent on valorizing distinctively feminine attributes and reclaiming a past in which women were powerful. A number of feminist archeologists object to the empirical (archeological) inaccuracy of these accounts and to the implausibility of their interpretive inferences. Conkey and Tringham review a number of these problems, noting the failure of most goddess theorists to take into account the diversity of the figurines they invoke as evidence of a prehistoric matriarchy and the complexity of the archeological assemblages from which these figurines are drawn. The figurines identified as goddesses are typically just one component of assemblages that include male and neuter, as well as female, figures and animal and artifactual, as well as human, figures. The extraction of goddess figurines from these assemblages and their assimilation to a prepatriarchal "goddess cult" obscure vast cultural differences between their cultures of origin, some of which are separated by several millennia of European prehistory and the whole geographical length and breadth of Europe. Conkey and Tringham argue that when feminists embrace these interpretations, they reinscribe in their own work precisely the gender essentialism—the stark oppositions and naturalizing assumptions about gender—that underpins sexist institutions and beliefs. Margaret W. Conkey and Ruth E. Tringham, "Archaeology and the Goddess: Exploring the Contours of Feminist Archaeology," in *Feminisms in the Academy*, ed. Donna C. Stanton and Abigail J. Stewart (Ann Arbor: University of Michigan Press, 1995), 199–247.

48. This parallels a response to the question of "how to do better" that has been developed in interesting philosophical detail by Helen E. Longino: "Can There Be a Feminist Science?" *Hypatia* 2, no. 3 (1987): 51–64.

49. For example, Kathryn Bernick, "A Stitch in Time: Recovering the Antiquity of a Coast Salish Basket Type," paper presented at the Annual Meetings of the Canadian Archaeological Association, Victoria, BC, 1998.

50. Marcia-Anne Dobres, "Gender and Prehistoric Technology," *World Archaeology* 27, no. 1 (1995): 25–49.

51. Diane Gifford-Gonzalez, "Gaps in Zooarchaeological Analyses of Butchery: Is Gender an Issue?" in *From Bones to Behavior: Ethnoarchaeological and Experimental Contributions to the Interpretation of Faunal Remains*, ed. Jean Hudson (Carbondale: Southern Illinois University Center for Archaeological Investigations, 1993), 181–99.

52. Cheryl Claassen, "Gender, Shellfishing, and the Shell Mound Archaic," in Gero and Conkey, *Engendering Archaeology*, 276–300.

53. Joan M. Gero, "Genderlithics: Women's Roles in Stone Tool Production," in Gero and Conkey, *Engendering Archaeology*, 163–93; and Joan M. Gero, "The Social World of Prehistoric Facts: Gender and Power in Prehistoric Research," in duCros and Smith, *Women in Archaeology*, 31–40.

54. Hastorf, "Gender, Space, and Food in Prehistory."

55. Mark N. Cohen and Sharon Bennett, "Skeletal Evidence for Sex Roles and Gender Hierarchies," in *Sex and Gender Hierarchies,* ed. Barbara D. Miller (Cambridge: Cambridge University Press, 1993), 273–96; Gillian R. Bentley, "How Did Prehistoric Women Bear 'Man the Hunter'? Reconstructing Fertility from the Archaeological Record," in Wright, *Gender and Archaeology,* 23–51.

56. Brumfiel, "Weaving and Cooking."

57. Sally Slocum, "Woman the Gatherer: Male Bias in Anthropology," in *Toward an Anthropology of Women,* ed. Rayna R. Reiter (New York: Monthly Review Press, 1975), 36–50.

58. Some early and influential contributions to the literature on "experimental archeology" and "ethnoarcheology" are John M. Coles, *Archaeology by Experiment* (New York: Charles Scribner, 1973); Richard A. Gould, ed., *Explorations in Ethnoarchaeology* (Albuquerque: University of New Mexico Press, 1978); and Carol Kramer, ed., *Ethnoarchaeology: Implications of Ethnology for Archaeology* (New York: Columbia University Press, 1979).

59. Hetty J. Brumbach and Robert Jarvenpa, "Ethnoarchaeology of Subsistence Space and Gender: A Subarctic Dene Case," *American Antiquity* 62 (1997): 414–36.

60. Spector, "Male/Female Task Differentiation."

61. Conkey and Gero, "Gender and Feminism in Archaeology," 415, 425.

62. Elizabeth M. Brumfiel, "Breaking and Entering the Ecosystem—Gender, Class, and Faction Steal the Show," *American Anthropologist* 94, no. 3 (1992): 551–67.

63. Ruth E. Tringham, "Households with Faces: The Challenge of Gender in Prehistoric Architectural Remains," in Gero and Conkey, *Engendering Archaeology,* 93–131; Ruth E. Tringham, "Archaeological Houses, Households, Housework, and the Home," in *The Home: Words, Interpretations, Meanings, and Environments,* ed. David N. Benjamin and David Stea (Aldershot, U.K.: Avebury Press, 1995), 76–107.

64. Gero and Conkey, *Engendering Archaeology.*

65. David Clarke, "Archaeology: The Loss of Innocence," *Antiquity* 47 (1973): 6–18.

66. Spector, *What This Awl Means.*

67. J. V. Romanowicz and Rita P. Wright, "Gendered Perspectives in the Classroom," in Wright, *Gender and Archaeology,* 199–223; Conkey and Tringham, "Archaeology and the Goddess"; Cheryl Claassen, "Workshop 2: Teaching and Seeing Gender," in Claassen, *Exploring Gender through Archaeology,* 137–53.

68. Spector, *What This Awl Means;* Tringham, "Households with Faces." Tringham is working with a team of researchers involved in the excavation of a Neolithic site in the former Yugoslavia, Opovo, to produce not only a conventional site report but also a video and a CD-ROM-based hypertext presentation of their data and interpretive hypotheses entitled *Chimera Web:* Ruth E. Tringham, "Multimedia Authoring and the Feminist Practice of Archaeology," in *Doing Archaeology as a Feminist* (Santa Fe, NM: School of American Research, 1998). Rosemary Joyce is coauthor, with Carolyn Guyer and Michael Joyce, of *Sister Stories,* a hypertext narrative based on the *Florentine Codex* (available on-line at http://www.nyupress.nyu.edu/sisterstor), as described in Rosemary A. Joyce, "Telling Stories: Doing Archaeology as a Feminist," in *Doing Archaeology as a Feminist.*

THREE

The Paradox of Feminist Primatology: The Goddess's Discipline?

LINDA MARIE FEDIGAN

How has primatology come to be beloved by many feminist science scholars? Has feminism truly helped to engender the field of primate studies as it exists today, or is this a case of mistaken attribution? Many science analysts have remarked favorably upon the feminist transformation of primate studies that has occurred over the past twenty-five years, and so widely accepted is the view of primatology as a feminist enterprise that Hilary Rose asks whether it has become "the goddess's discipline."[1] Donna Haraway, in her influential analysis of the history of primatology, and the authors of lesser known analyses (e.g., Rosser's application of the "six stages of feminist transformation" to primatology) have argued for a shift in primatology toward the values and practices of feminist science.[2] But what makes this case curious and paradoxical is that most primatologists vehemently deny that theirs is a feminist science. Only a small handful of primatologists are self-declared feminists, albeit scholars whose work has been very influential—Jeanne Altmann, Sarah Hrdy, Jane Lancaster, Barbara Smuts, and Meredith Small, for example. And one need only peruse a few of the strongly negative reviews by practicing scientists of Haraway's book on primatology ("infuriating" is a common reaction)[3] or attempt to casually interject the word "feminism" into a discussion with primatologists or ask them outright if they consider themselves feminists to uncover the strength and depth of their denial. In this chapter I explore how primatologists have responded to the feminist critique of science by becoming more gender inclusive and by using "tools of gender analysis,"[4] and I suggest why so many primatologists carry out work that closely adheres to the tenets of feminism while at the same time denying the appellation. I argue that this contradiction is much more than a "label problem"—rather that these primatologists see themselves as operating from vastly different underlying assumptions and models of science than those of feminists.

An attempt to understand the paradox of feminist primatology is important for several reasons: it offers insights to those who want to know how and why a science changes over time; it can help to establish better communication and working links between scientists and science analysts; it exemplifies one discipline's solution to the "women in science problem"; and it sheds some light on the central issue of this volume: What useful changes has feminism brought to science? It does take some audacity for me to even try to address the last issue since many will feel that Donna Haraway definitively answered the question of how feminism has affected primatology in her comprehensive and multifaceted analysis of the subject.[5] But Haraway herself would surely advocate multiple voices and diverse perspectives on this topic. And I have puzzled for some time over the deeply negative reactions of my fellow primatologists to Haraway's interpretation of primatology as a "genre of feminist theory," and over the extent to which some of my scientific colleagues mistrust the label of feminism even while practicing feminist virtues. I have also pondered the question of what type of evidence would convince scientists that feminism has brought useful changes to their practices. I cannot claim to have definitive answers, but perhaps I can bring to bear on this matter the multiply "situated knowledges" of a practicing primatologist, a feminist, and an interested reader of science studies.

Both Haraway and Hrdy have argued that it is not simply a coincidence that primatology in North America began to develop along feminist lines in the 1970s at the same time that the second wave of the women's movement was cresting in American society.[6] Primatology is a quantitative and empirical science, and thus, for many practitioners, the most convincing evidence of the impact of feminism on primate studies would be to demonstrate directly through experiment or observation that the feminist critique of science can cause (or has caused) primatologists to change their practices. It would also be convincing if a substantial proportion of primatologists stated that their work has been influenced by feminism. It seems to me that neither of these scenarios is likely to be enacted and that we will therefore have to rely on indirect evidence. In a previous paper, I decided to set aside the issue of which primatologists self-identify as feminists and asked instead, What would a feminist science look like? and Does primatology look like this?[7] I identified six features that many models of a feminist science hold in common (reflexivity; taking the female point of view;

cooperation with, rather than domination of, nature; moving away from dualisms and reductionism; humanitarian applications of science; and greater inclusiveness of formerly marginalized groups) and argued that the field of primatology has increasingly exhibited these features over the past twenty-five years. From this circumstantial case, I concluded that primate studies can be called a feminist science or that, at the very least, it has been significantly changed by the women's movement and the feminist critique of science.

Since writing that paper, I have come to think that what is unusual and instructive about the case of primatology is not that it can (arguably) be labeled a feminist science but rather the rapidity and the extent to which it "self-corrected" in response to the feminist critique of science. By this I mean that many disciplines have been the subjects of feminist critiques (e.g., archeology, biology, cultural anthropology, history, physics), but few, if any, others moved so quickly, so extensively, and so willingly to rectify the previous androcentric aspects of their practices. Therefore, in this paper I want to bypass the probably irresolvable debate as to whether or not primatology is a feminist science and instead begin to detail some of the myriad ways in which primatologists have become increasingly gender sensitive and gender inclusive over the past twenty-five years. I will attempt this documentation through the heuristic device of what Londa Schiebinger has called "tools of gender analysis." Schiebinger has advocated a move away from prescriptive visions of feminist science and suggested that we instead highlight a set of tools of gender analysis that have been used in many different sciences for designing woman-friendly research along feminist lines—tools that do not "create some special, esoteric 'feminist' science, but rather . . . incorporate a critical awareness of gender into the basic training of young scientists and the work-a-day world of science."[8] Schiebinger cautions that these analytic devices are not peculiar to feminists, and I will provide many examples of primatologists using these devices—some of whom do, and some of whom do not, call themselves feminists. Then I will return to the paradox laid out in this introduction and suggest several reasons why many primatologists carry out research along feminist lines even while they distance themselves from any such association.

EIGHT TOOLS OF GENDER ANALYSIS
Tool 1: Scientific Priorities

Schiebinger argues that one of the most important gender analytics looks at scientific priorities—given limited resources, how are choices

made about what we want to know and what we will fund for researchers to study?[9] The funding world of primatology is dominated
by issues that granting-agency review panels believe to be of grave
concern to human society. Almost every young primatologist must
learn to answer the question, How is your research relevant to humans? That is one of the reasons why Haraway chose to analyze primatology as a science—because it intersects with so many vital human
social issues, for example, mother love and male aggression, colonial
and postcolonial impact on the tropical nations where primates occur,
race, class, sex, politics, war, peacemaking, conservation, views of nature, views of human nature, and the list goes on.[10] Many people approach the study of our primate relatives as if these animals will provide us with nature's true blueprint for human nature. Primatology
thus becomes what Haraway variously calls a "trading zone" or "zone
of implosion" where researchers from many backgrounds vie for high
stakes—the right to propound their vision of how we humans were
meant to behave and to live together.[11] And in this context what we
choose to study, how we study it, and what we choose not to study
become crucial.

From the 1930s until the mid-1970s, it was quite common in North
American and European primatology to consider females only in the
role of mothers and as passive resources for males.[12] The publication
of the prescient article by Jane Lancaster entitled "In Praise of the
Achieving Female Monkey" heralded the beginning of a new era.[13]
Primatologists began to rescue the "hidden females" from their shad-
owy, secondary roles and to bring them onto center stage. Some
researchers, such as Jeanne Altmann, Sarah Hrdy, Thelma Rowell,
Meredith Small, and Barbara Smuts, have talked about how they deliberately chose to factor sex differences into their studies, as well as to
study females, and to take the "female point of view" as part of their
feminist approach; many others have done so without commenting on
what thinking lay behind the process.[14]

I would argue that there are at least three ways in which primatologists over the past twenty-five years have deliberately shifted their priorities about what we study in relation to gender issues. One fundamental shift has been simply to provide more funding for research that
fleshes out the picture of what female primates do, apart from bearing
and rearing offspring—the patterns that are part of what Jeanne Altmann has called "dual career mothering."[15] This shift paralleled the
drive in many other fields to ensure that women formed part of the
study sample, for example, in psychology and medical research. A second early change in primatological priorities was to turn standard re-

search questions around and ask them from the female's perspective. One telling example comes from the study of size and shape differences between males and females of the same species. For over a hundred years, since Darwin first posed the question in print in 1871, evolutionary biologists have puzzled over sexual dimorphism and asked themselves, Why are male mammals usually bigger than the females of the species? It was not until the late 1970s that scientists began to ask, What might be the adaptive advantages to female mammals of being smaller than the males of their species?[16] This simple inversion of the question changed our entire perspective on the conundrum—we went from talking about how males need to be large in order to protect females to examining the different bioenergetic demands on female and male mammalian bodies.

But the third, and probably the most telling, example of how primatologists have shifted their priorities about the types of research questions that are asked and funded is the increasing study in the 1980s and 1990s of gender-related issues that clearly concern women when they arise in the human context. For example, Smuts has published both empirical research and theoretical papers on topics such as male aggression and sexual coercion of females, male-female friendships, and male dominance and its repercussions for females.[17] Hrdy has studied why male primates sometimes kill infants, how females compete with other females, and why parents favor sons over daughters.[18] Lisa Rose and I embarked on a series of papers addressing the issue of why female primates live with males on a year-round basis—examining what benefits males provide and in what ways males are a liability for females.[19] Small has published books about women primatologists studying female primates, about the processes by which females choose their sexual partners, and about the evolution of female sexuality.[20] Lancaster has published extensively about parental care and its evolution in the primate order and about the evolutionary biology of women.[21] There are many more examples, but these should suffice to show that we have created a new vision of the female primate in large part by simply choosing to find out more about her and her relations with the others of her species from her own perspective.

Tool 2: Representative Sampling

As pointed out by Schiebinger, one of the basic tools of gender analysis is to make scientists aware of the appropriate inclusion of females as subjects of research and of women as the conductors of research.[22] To some extent, I covered the issue of females as subjects of research above, but here I will briefly describe two ways that sampling patterns have been recognized by primatologists to have implications for the conclusions we draw about sex and gender.

The first example is the landmark and now canonical 1974 paper by the primatologist Jeanne Altmann on appropriate methods for sampling animal behavior.[23] Although she is sometimes mistakenly credited for having invented these sampling methods, what Altmann did was to codify, clarify, and label the methods used by ethologists to sample behavior up to that time and to authoritatively evaluate their relative strengths and weaknesses for different research designs. The influence of this paper is enormous: it standardized sampling practices in ethology and because it is still referenced in most methods sections of ethological publications, some credit it with being the most cited paper in the modern literature on animal behavior.[24] It also discredited for most purposes what is called "ad lib sampling"—the practice of opportunistically recording whatever strikes the observer's eye and gains one's attention. In particular, Altmann established that it is inappropriate to use such opportunistic sampling to compare rates of behavior between individual subjects or between males and females, for example. What this meant for primatologists is that we stopped watching only the larger, more swashbuckling males and started to also sample for representative periods of time the less prepossessing subordinate males, the females, and the immature individuals of the groups. Altmann made no mention in this famous publication that she was dissatisfied with the previous bias toward observing male primates more than females—rather, what she did was to convince scientists, through the use of their own methodological tools, to raise their standards of evidence, a task for which she was well qualified by her background training as a mathematician. Elsewhere, she did state that her involvement in the feminist movement contributed to her awareness of and dissatisfaction with previous androcentric sampling practices.[25]

Another way in which primatologists realized that sampling can have implications for gender is the process by which we choose which species to study. For example, out of approximately two hundred primate species, the male-dominant savannah baboon was long the favorite species for modeling the evolution of human life.[26] In the 1950s and 1960s, a large contingent of North American primatologists, influenced by Sherwood Washburn, believed that we should be able to document something they called the "primate pattern," a code term for the basic "nature" of primates.[27] Aspects of what was then presented as baboon behavior, such as male bonding and aggression and rigid male dominance hierarchies, were generalized to all nonhuman primate species as representing the primate pattern, and from there to the human species as well. The first criticisms of this "baboonization" of primatology came as early as the 1970s.[28] Initially, the baboons were replaced by other single-species models, such as the chimpanzee model for hominid evolution, but today it is realized that no one species will provide a sufficient model for early human behavior. Although primatologists are now largely aware of the biases introduced when we generalize from one (or even a few) species to the entire order, we still make many assumptions about primate nature on the basis of the species we have studied (mainly Old World monkeys and apes), and we are finding these assumptions challenged as new data roll in on the less well studied species.[29] With so many species of primates, there are examples of almost every type of social system and relationship between the sexes that one could think of, and it makes a great deal of difference to our picture of human evolution and gender relations which species we choose to study and to emphasize.

Tool 3: Dangers of Extrapolating Research Models
from One Group to Another

Schiebinger gives examples of the dangers of extrapolating to women research models designed for men.[30] There are some parallel examples in primatology, such as the study of sexual dimorphism mentioned earlier. One of the stumbling blocks that hindered scientists from asking why female mammals might be selected to be smaller than males was their initial neglect of the repercussions of gestation and lactation. In primatology, it was sometimes assumed that the principles that apply to body size differences between species, such as those between chimpanzees and gorillas, could be applied willy-nilly to body size differences between the males and females of a species. Thus, it was not at all uncommon in the early days of primate ecology studies to read that the larger males of dimorphic species need to eat more and use a larger share of resources than do the females. However, a female primate spends most of her adult life gestating or lactating and may have feeding and metabolic rates up to 200 percent greater than a nonreproducing female of the same size. Today, primatologists are very aware of the different physiological, reproductive, and life history processes of male and female animals and much more careful about extrapolations from one sex to the other. If anything, primatologists sometimes follow the lead of those evolutionary theorists who suggest that the males and females of a given species can be so different in their biology and behavior as to be almost "two different species."

But where I think that primatologists have become the most aware of the dangers of extrapolation from one group to another is in the projection of Western gender role stereotypes onto animal patterns and onto our human ancestors. We can see this projection most clearly in some of the Victorian scenarios that nineteenth-century Eurameri-can anthropologists and natural historians imagined for monkeys and for early human social life, in which females were described as domes-tic, reluctant, and coy, whereas males were thought to be assertive, competitive, and mobile. But even in the 1960s and 1970s, when field reports on primates were first published, we began with many assump-tions specific to the gender relations of our own time and place. For example, male monkeys and apes were sometimes referred to as "own-ing" their females,[31] whereas we know today that males in many of these species merely attach themselves on a temporary and rotating basis to permanent kin-bonded groups of females who occupy a consis-tent matrilineal home range. Can it be a coincidence that a generation raised on *Father Knows Best* would have propounded the 1950s view that primate females were mothers and mates and did little else of social or ecological note? How else to explain the influence of and the vast number of papers published on the "priority of access model," which assumed that female monkeys were passive resources available to males as sexual partners simply in order of the "winner's list" from male-male competitions? One by one these assumptions have disinte-grated as researchers have focused on the females and documented not only the active roles they play in primate society but the enormous variety of relations between the sexes that occur in primates, defying our attempts to simplify and extrapolate to one "human nature." And the important point for our purposes here is that ethologists and pri-matologists themselves became aware of the dangers of extrapolating from humans to animals and then back to humans. Several researchers published warnings about the rebounding anthropomorphism known as the "aha! reaction" (e.g., "Aha! Female animals are choosy about mates; therefore, we can refer to them as 'coy.'" "Aha! Human females are coy just like animals; therefore, female coyness must be part of our primate heritage.").[32]

Tool 4: Institutional Arrangements

Schiebinger argues that institutional power structures influence the knowledge that issues from them—from formalized universities to recognized-but-informal schools of thought to nearly invisible "cliques." The examination of how such institutions structure scien-tific representations of gender, race, and nature was of course the pri-mary tool used by Haraway in her extensive analysis of primate stud-ies.[33] Use of this tool does require the assumption that "primatology

is politics," and this may be one of the reasons that few, if any, primatologists themselves have commented in print on how institutional
background affects what we believe to be true about primates and
about gender. Furthermore, the scrutiny of institutional power structures may require a more uninvolved perspective on the science than
many scientists feel they are ready or qualified to take. One exception
is Hrdy's brief analysis of the relationships among several important
variables in her career: her status as the sole woman in her cohort
at Harvard graduate school in the 1970s, her dawning awareness of
androcentric bias in animal behavior studies, and her growing ability
to imagine female animals as active strategists.[34] A second exception
is a recent Wenner-Gren–sponsored workshop of primatologists and
science analysts in which many participants concluded that institutional background is even more important in influencing a primatologist's thinking than is theoretical affiliation or methodological preference.[35]

Tool 5: Gender Dynamics in the Cultures of Science

I teach in a building of social science departments but attend many
meetings in the neighboring biological sciences building. Perhaps because of my background training in sociocultural anthropology, I have
often noticed the radically different cultures reflected in the dress codes
apparent in these two buildings: suits and pantyhose in the social science building; field clothes and lab coats next door. I also note differences in the intellectual styles and acceptable behavioral patterns of
the biologists and the social scientists as I commute between them
daily. This has clear implications for women graduate students in primatology, who may major in either anthropology, biology, or psychology, and who then face a choice between becoming "one of the boys"
in the biological field or lab situation or taking on the "woman in
power suit with briefcase" role in a social science setting. Primatologists have not commented in print about the "culture of primatology,"
perhaps for the same reasons that they have not reflected publicly upon
institutional arrangements. But its "scientific culture" is one of the several ways in which primatology is a very informative case study, because I believe that ours is by and large an androgynous culture. To
some extent, this is due to critical mass—there are now nearly equal
proportions of men and women primatologists in North America, and
there have always been substantial proportions of women in this science.[36] The most telling example of our androgynous culture is the
nature of fieldwork, which is strongly gendered masculine in most sciences (e.g., archeology, paleontology, geology, entomology) but which

is not gendered masculine or feminine in primatology. For reasons which are beyond the scope of this paper (e.g., the relatively recent emergence of primate fieldwork after World War II, and the early role models of women anthropological field-workers), a macho fieldwork image never took hold in the science of primatology in North America (Japan, Latin America, and India are different stories). It has always been expected that women will be as able and as competent as men to carry out field research on primates—research that is just as physically demanding, if not more so, than it is in other sciences. In fact, the popular media's attraction to women primatologists "roughing it" has sometimes given the public the impression that only women do field-work in primatology, but this is not the case. There are many successful men working at primate field sites, although it is possible that the men are more associated with short-term studies and the women with longi-tudinal field studies (this would make a good research question). My point is that I do not think in this case that feminism changed the gendered nature of primate fieldwork; I think rather that women got in on the ground floor, proved early on that they could do good work in the field as well as in the lab, and helped to establish an androgynous culture of primate science in which feminist tenets could flourish.

Tool 6: Language Use

Language both reflects and structures our thinking, and as noted by Schiebinger, much gender analysis has focused on the rhetoric of scien-tific writings and speech.[37] Analysis of gender symbolism in scientific terminology has been a particularly effective tool for feminist critiques of science because most scientists are strongly encouraged to be clear and precise in their language. However, they may not be fully aware of the powerful connotations of the metaphors and other figures of speech that are embedded in all language because this is not part of their formal training. Thus, it is probably easier to convince a scientist of the androcentrism inherent in references to "passive" eggs and "ac-tive" sperm (since it is now obvious to any biologist that these are not appropriate descriptors) than it is to convince scientists that institu-tional power structures affect the knowledge that issues from them or that there are gender dynamics in the culture of science.

Primatologists are caught on the horns of a powerful language di-lemma in the form of anthropomorphism. One of the strongest taboos in primate studies is to attribute human characteristics to animals—before we have studied them to determine how they do, in fact, behave and think—since this implies that all organisms behave and think like ourselves. And yet, we cannot fully invent a new language to describe our observations of animals, so we must borrow terms from the human domain that seem to best capture what we observe in the behavior of our animal subjects. The more closely related the animal is to us, the more it looks like and seems to behave like us, and thus the greater the danger of anthropomorphism. Primates are our closest relatives and share many characteristics with humans. But careful observations by scientists have convinced us that the gorilla does not beat his chest because he is angry; the vervet does not present his posterior to the face of another vervet to insult him, and the macaque does not grin because she is happy. Anthropomorphic assumptions about the meaning of primate signals must be avoided at all costs in that basic tool of all primate behavior research—the ethogram, which is a list of the behavioral units to be studied. Much of the work of developing a good ethogram involves language use, and a considerable amount of time and training is invested in teaching students to use descriptors that are as value neutral as possible (e.g., "open-mouth gape" instead of "threat face").

Thus, primatologists are somewhat predisposed to be judicious in their use of language. However, this is not to say that there have not been biases in the choice of terms to describe primate behavior. Biased terminology was a sufficient problem to warrant my devoting an entire chapter to language use in my general review of primate sex roles and social bonds.[38] In the early 1980s, I identified two major forms of biases in the choice of terms to describe primate behavior: androcentrism and a preference for hostile and combative metaphors. I was certainly not the first or the only scientist to identify sexism in the language of primate behavior.[39] Perhaps a single example will suffice. One type of social organization found in some primate species consists of several adult females who rear their young together in the presence of one adult male at a time (a uni-male, multifemale, or polygynous, system). This social system had been described by Darwin (1871) as "harems" ruled by "despots" or "masters" who "possessed" the females, and some of the early primate field studies reiterated such language.[40] However, after a few years of study it became obvious that in almost all polygynous primate societies (hamadryas baboons being the exception), a network of related females forms the stable core of the group and males come and go—hardly our idea of a "harem." Once it was recognized that these sorts of terms were very inappropriate to the actual behavior of most primate species, they were largely dropped. However, the hostile and combative metaphors have been harder to discourage. I wonder what comment it makes on our scientific culture that anthropomorphic and value-laden terminology that connotes sentimentality or "a warm fuzzy feeling" (e.g., "aunting" rather than "allomothering"; "kids" rather than "juveniles"; "babies" rather than "infants") has been mainly rejected whereas equally anthropomorphic terms that imply belligerent or ethically undesirable imagery in the human context (e.g., rape, cheaters, suckers, selfish) are still widely used and accepted.

One more area of language use that has changed in primate studies concerns the active/passive connotations of how we describe the behavior of the animals. For example, there used to be two terms to describe the sexuality of female animals: attractiveness (is she attractive to the male?) and receptivity (does she accept male advances?). In other words, the female was seen as a passive resource for males. In 1976, Frank Beach pointed out that estrous female mammals often take the initiative in approaching, investigating, and soliciting sex. He called this phenomenon "proceptivity," and it has since come to be extensively used for and documented in female primates.[41]

Haraway identified the granting of agency as an important component of the situated knowledges of feminists and pointed out that primatologists (mainly but not only women primatologists) have activated the previously passive category of the female body and of female sex.[42] This activation of females is part of a larger move in primatology and anthropology to depict the "object of study" as an actor or agent rather than as a passive resource. In primatology, the granting of agency to the animals has been part of two larger moves: the modernization of sexual selection theory and the cognitive revolution. The theory of sexual selection developed by Darwin over a century ago consists of two main principles: male-male competition and female choice.[43] Although male-male competition has been studied over the past hundred years in great detail in many species, female choice of mates (and indeed male choice of mates and female-female competition) has only been well studied over the past couple of decades. Until recently, scientists have shown little enthusiasm for Darwin's idea that female animals could act as selective forces through the careful choice of their male mates and thus influence the evolutionary direction of modifications in male appearance and behavior.[44] Similarly, scientists used to be reluctant to consider what goes on in the minds of animals and, under the influence of the theory of "behaviorism," assumed that animals mainly reacted to their environment without mental reflection. Therefore, primate society used to be depicted as very much structured by "brute force": the biggest, strongest individuals ran the show, the alpha male got all the females and produced all the infants, and so on. Then, as part of the cognitive revolution, primatologists began to investigate the possibility that primates think before they act, that they are sentient creatures who are able to remember past events, to recognize relationships among others along kinship and dominance lines, and to predict the consequences of their actions and those of others. This new granting of agency to primates is reflected in the language that is now used to describe their behavior—"social strategies," "skilled tacticians," "primate politics" (the ability to "finesse" instead of fight one's way to success), "deception," "peacemaking," and so forth. Recognizing the cognitive abilities of our fellow primates, especially their strategic skills, gave us an explanation for the widespread finding from field studies that the most influential individual in the group is not usually the largest or strongest male. It may also help to explain the startling findings from DNA paternity studies that infants born to ape and monkey groups in the wild are not always fathered by the males of the group,[45] even when the females appear to be subordinate to these males and mating only with them.

Tool 7: The Remaking of Theoretical Understandings

Schiebinger notes that there has been controversy about how deep gender analysis goes and whether feminists have contributed to the remaking of the theoretical understandings of a discipline or just "added females" to the mix.[46] The idea that in scholarship women are the fact gatherers and men are the theoreticians seems to be pervasive although seldom stated in print or studied directly. In her consideration of the gender of theory, Catherine Lutz has argued that the lines separating theory from nontheory are fuzzy, that theory is intentionally or unintentionally signaled to and picked up by readers as more associated with male scholarship than female scholarship, and that feminists have continually pressed against the dualism of theory and practice.[47] This latter point was brought home to me by Naomi Quinn, who referred to some of my own work as theoretical when I thought of it as merely a critique. As she pointed out to me, "How does any new theory arise except in the context of other theory?"

One of the "signals" of theory is that it is more abstract, original, and generalized than other writings and that it articulates thoughts not spoken or published before. Bruno Latour put this last characteristic of theory somewhat differently. He argued that it is important to document the "interesting differences" made by women scientists, differences that have shown primates to us in a new light or allowed primates to "speak" to us in new ways.[48] Using this understanding of theory, I would like to address some specific cases where primatologists have been instrumental in allowing female primates to become significant social actors or where they have created new "setups" (to use another Latour term) that changed our perceptions of female primates. In the field of primatology, there have been many such important theoretical breakthroughs, brought about by the ideas of both women and men. Some of these I have briefly described already or Haraway has valorized in her widely read analysis of primatology; others are lesser known.[49] Although I recognize the many pitfalls of providing a simplified list of important theoretical insights first developed by individual primatologists, in the interest of space (and in the interest of including mention of people who are not well known outside their science) I offer here brief synopses of some of the important work by primatologists that has remade our theoretical understandings of female animals and the relations between these females and the males with which they live. This list is neither exhaustive nor representatively sampled; it is rather what comes to my mind when I think of theoretical advances that changed our understanding of female primates.

1. Several researchers contributed new insights about principles of baboon behavior that re-created the world of social baboons (and thus the generalizations about all social primates) from a male militaristic model to a representation in which females as well as males play active roles.[50] For example, Altmann conceptualized and documented female baboons as "dual career mothers" with the capacity to feed themselves and their nursing offspring, to form long-lasting kinship bonds, to act as repositories of ecological knowledge, as well as to rear their young.[51] Thelma Rowell spearheaded the critique of the "male dominance" model of baboon social life and argued that dominance may be more characteristic of human primatologists than it is of the nonhuman primates, and that its common appearance in baboons and macaques might be the result of human-induced conditions in captivity.[52] Shirley Strum documented how individuals with real power are those who can mobilize allies rather than those who can push through with brute force; they rely on systems of social reciprocity, which they actively construct.[53] Barbara Smuts developed a model of "friendship" in baboons that showed how adult males slowly become accepted members of matrilineal societies by ingratiating themselves with individual females and their infants.[54]

2. Richard Wrangham modeled primate society as one where females first distribute themselves according to the resources (food and water) available, and males then distribute themselves according to the spatial and social pattern of females available.[55] This idea was enor-

mously influential in moving us away from a concept of male primates as "owners" of females, because Wrangham's model causes us to think of female patterns of distribution and female sociality as being prior to that of males.

3. Donald Sade recognized from his long-term studies of rhesus monkeys that female kinship bonds are the fundamental structuring principle of most Old World monkey societies. Although many others contributed to this important theoretical understanding (especially Japanese primatologists), Sade published several influential papers exploring the implications of his findings that affiliative interactions occur mainly among matrilineal kin, that mothers avoid mating with sons, and that groups fission along kinship lines.[56]

4. Sarah Hrdy is well known outside the discipline of primatology for her theoretical contributions to sociobiology.[57] She brought female strategies and a female perspective into sociobiological theory, she reformulated female primates as active strategists and competitors, and she conceptualized female behavior as adaptive strategy. In her work on infanticide she has both developed and tested the hypothesis that adult males kill infants fathered by other males as an adaptive strategy to promote their own reproductive success.[58] Although Hrdy considers herself a feminist sociobiologist, the theories of sociobiology have been distasteful to many feminists, which has given rise to the contradictory treatment she has received in the feminist science literature.[59]

5. Michael Huffman, Charles Janson, Joseph Manson, and Meredith Small examined the ways in which female primates choose their mates and express their preferences.[60] Small in particular has developed generalizations about the principles by which mate choice operates in primates. As noted above, scientists took nearly one hundred years to get around to documenting this essential principle of sexual selection theory first proposed by Darwin, but since then the study of female choice has been a key factor in remaking our theoretical understanding of relations between the sexes.

6. Karen Strier showed us that contrary to the popular image of the aggressive, competitive male primate, many New World monkey societies (primarily those of the atelines) include philopatric, affiliative, closely bonded adult males, as well as females who transfer between groups.[61] Strier has been at the forefront of challenges to the generalizations we have all developed about primates based only on the well-studied Old World primates, especially the baboons, macaques, and chimpanzees, and she has encouraged us to acknowledge and incorporate into our theories the many forms that male and female relationships may take in primate societies.[62]

7. Several other important insights into how primate societies function have come from the more recent studies of New World monkeys. For example, group movement patterns are a useful indicator of social dynamics, coordination, and power. Many primatologists have assumed that dominant males determine the direction of group movement, although it is usually difficult to tell who is leading whom. Sue Boinski demonstrated that high-ranking adult female capuchins use a particular vocalization and body stance to draw the group's attention to the intended direction of her travel, and the group then moves in the same direction as the signaler.[63] This suggests that it may not be the animal out in front, but the one giving a particular set of signals, and the ones with greater historical knowledge of group ranges (in this case, the philopatric females), who are influencing group movement.

8. Carolyn Crockett, Kenneth Glander, and Margaret Clarke have shown that in societies of New World monkeys such as howlers where adolescent females transfer between groups, and therefore do not have kinship bonds with the other adult females of their groups, female-female competition can be fierce to the point of fatal wounding.[64] This challenges the stereotypical view of female primates as always being the affiliative forces in society.

9. Devra Kleiman and Patricia Wright drew our attention to the extensive parental care shown by adult males in monogamous primate societies and developed explanations for the conditions under which male care and monogamy are adaptive in primates.[65]

10. The more recent studies of prosimian primates have also challenged some of our assumptions based only on the Old World monkeys and apes. For example, Alison Jolly, Michael Pereira et al., and Alison Richard brought dominant female lemurs to our attention and helped to develop an evolutionary and ecological model of the conditions under which it is advantageous for female primates to be dominant over the males of their groups.[66]

Although there are other examples of how our theoretical understandings of female primates and of male-female relationships have been remade, I hope that these ten will suffice to show that primatologists have been very concerned to provide a more complete picture and a greater critical awareness of the roles of females in primate societies and to disassemble old sex role stereotypes.

Tool 8: Challenges to What "Counts" as Science

It is fairly common in the feminist literature to point out that bodies of knowledge more associated with women are often categorized as "nonscientific"—home economics, family studies, and nursing, for example. Schiebinger argues that gender analysis has challenged what "counts" as science.[67] I would like to briefly describe two areas in which some Western primatologists, mainly women, have challenged what counts as primatological science: empathy as means of understanding the animals and "mission" science.

Empathy is the projection of one's own feelings and thoughts onto the emotions and behavior of another and thus is quite similar to anthropomorphism, the primary taboo in primate studies, which I have already briefly discussed. Although few Euramerican primatologists have dared to suggest that they employ empathy to better understand their subject matter, the exceptions that I can think of have almost all been women (several of them also on our list of self-declared feminists): Hrdy, Rowell, Sicotte and Nisan, Small, and Strum.[68] The primary exception to this gender coding of empathy as feminine in North American primatology comes from the senior and respected animal behavior field-worker George Schaller, who said that much of what we understand about primates we do through intelligent empathy.[69] Another exception comes from Japanese primatology, which developed simultaneously and independently of Western primate studies and in which most practitioners are men who use "empathetic understanding" as a scientific tool.[70] Empathy is a two-edged sword for women primatologists in North America and Europe because their hard work can easily be dismissed as some form of "female intuition," and because admitting to empathy can jeopardize one's reputation as an objective observer. It seems that not only is it taboo to employ empathy as part of one's scientific tool kit, but most scientists are even reluctant to talk about it in print or to analyze the roles that it might play in our work. At the moment, it is certainly not thought to count as science outside Japan.

Mission science in primatology mainly takes the form of conservation work. Although many men (Russ Mittermeier being the most famous)[71] and indeed most field primatologists are necessarily involved to some degree in conservation of the species they study, I would like to consider for the purposes of this chapter the example made famous by the media: the "trimates," Jane Goodall, Dian Fossey, and Birute Galdikas. There are other more in-depth analyses of these three women and their work,[72] but I want to focus here on only two aspects: their perseverance and their willingness to sacrifice scientific success in order to devote themselves to conserving and enhancing the living conditions of the great ape species that they have worked with for so long. Obviously, other primatologists have also carried out longitudinal research, but the fact that Jane Goodall is still overseeing field research on the chimpanzees of Gombe forty years after she first began is truly remarkable. Dian Fossey's research was of course cut short by her untimely death, and Birute Galdikas has run into problems renewing her permit to carry out research at her field site, but her students and field assistants continue it for her. Each of these women decided at some point in their long careers that the need to save the endangered species they study is greater than the need to save their reputation as pure and productive scientists, and each has suffered from the diminished respect that some scientists accord to popularizers of science and to those who break ranks by placing their politics on a par with, or above, their science. I do not wish to turn these women into saints or martyrs, for that they certainly are not. However, Haraway's description of Jane Goodall as the "Virgin Priestess in the Temple of Science" certainly seems apt when one sees the masses of people lined up for a simple touch of her hand.[73] Not only have these three women primatologists triggered a major tide of public goodwill and funding for primates and the scientists who study them, but their example has also encouraged many young people, particularly young women, to enter this discipline. And they have played an important role in making conservation and animal welfare a very active and recognized part of the science of primatology. Conservation symposia are now the most widely attended sessions at national and international conferences of primatology, and the *International Journal of Primatology* flags with a special symbol all articles published on endangered species. I would argue that the courage of Goodall, Fossey, and Galdikas to break ranks and their challenge to what counts as science have been crucial in legitimizing mission science as an accepted aspect of primatology.

FEMINIST PRIMATOLOGY—WHAT'S IN A NAME?

In the previous section, I have provided numerous examples of primatologists using the tools of gender analysis identified by Schiebinger as methods commonly employed by feminists to help create woman-friendly science. The literature on gender and science indicates that there are at least three fundamental ways in which feminism can influence a science: (1) it can create more opportunities for women to enter and succeed in science (which has no further implications for change in science if women "do science" just like men); (2) it can increase gender awareness or gender sensitivity in the practitioners of a science;[74] and (3) it can alter the working dynamics of a science (e.g., power relations, gender symbolism) through the practices of well-established scientists who are also feminists. I would argue that primatology has exhibited changes over the past fifty years of its existence that provide substantial evidence for all three types of influence. First of all, our discipline has included higher and higher proportions of women practitioners,[75] and these women are holding more of the important societal offices of our profession (e.g., a recent past president of the International Primatological Society was Alison Jolly, and there have also been women presidents of the American Primatological Society). Second, the many examples that I provided above of primatologists changing their minds, their descriptions, and their research foci to recognize and document the "achieving female primate" are in part the result of increasing gender awareness in primatologists. Few North American primatologists today would refer to a polygynous society as a "harem" or treat females as if they were passive resources for male competition. Third, I have given examples of the few but powerful women primatologists who identify themselves as feminists, such as Sarah Hrdy and Jeanne Altmann, and who have played significant roles in rectifying the early androcentric biases of our field.

Why then do so many primatologists distance themselves from any association with feminism and deny that they have been influenced by feminism? I will offer three possible reasons based on my knowledge of, and conversations with, primatologists. First, I think that there is an underlying concern that our science not be perceived as "feminized," because of the widespread belief that feminized disciplines become devalued.[76] Most scientists do not perceive a difference between a feminine science and a feminist science and so would confuse the influences of feminism on science with "doing science in a feminine way" (i.e., "science with pink ribbons" as my colleague Shirley Strum puts it). Primatology is a quantitative science with a vital component of evolutionary biology and a strong association with other fields coded "masculine," such as anatomy, physiology, neurology, paleontology, and quantitative ecology. Training for both lab and fieldwork is rigorous, and many primatologists feel that all the media attention to women primatologists holding baby apes will project an incorrect image of a "sissy science."

Second, like scientists everywhere, primatologists distance themselves from anything perceived as political, because "politics" implies bias and failure to adhere to the scientific credo of objectivity. As Schiebinger has pointed out, feminism is a dirty word in North America and has many negative connotations to scientists.[77] Much of the science studies literature refers back to C. P. Snow's classic description of the "two cultures" of science and humanism, two cultures that still hold sway and are dichotomized in the minds of scientists.[78] Most of the primatologists I know see themselves as scientists, whereas they classify feminists as not-scientists. They are at least vaguely aware of the "science wars" and often assume that feminists must be on the other side. An interesting point, worthy of study, is how and why scientists classify certain theories as political (e.g., any feminist theory) and other theories as nonpolitical (e.g., sociobiology). One primatologist explained this distinction by saying that he does not carry his sociobiological beliefs home to the bedroom, whereas his wife does carry her feminist theories home. I suspect that the distinction between political and apolitical theory is often made on the basis of where the theory is thought to have originated—"in" or "out" of science. Since sociobiology was developed by scientists, it is perceived as nonpolitical in spite of its strong implications for, and widespread adoption by, conservative political groups in the United States. Feminism is perceived by primatologists as having been developed outside science, in fact as a critique of science, in spite of the work of influential and established feminist primatologists such as Sarah Hrdy, who is one of the few women (and perhaps the only feminist?) to be elected to the National Academy of Sciences.

Finally, many of the primatologists that I know hold a very idealistic view of science, a view that we inculcate, both consciously and unconsciously, in our students during their graduate training. Such a view has been called the "Legend of Science" by Steven Shapin and "Science with a Capital S" or "Science-Already-Made" by Bruno Latour.[79] Holders of such an idealistic view tend to perceive science as operating in a different realm from all other human activities, a realm that is pure and objective and free from sociocultural influences. The messiness of the actual practice of primatology in the field and in the laboratory is not denied so much as it is de-emphasized or ignored. That the community of scientists insists on Science with a Capital S is exemplified in the "cleaning up" of what we report in our publications and grant proposals—our research would never be published or funded if we laid out all the influences, all the mistakes, all the blind alleys, and all the human controversy that are, in fact, an integral part of the work.

Many scientists see themselves as operating from a nonpartisan position, and they see feminists and other science analysts, who often focus on the untidiness of "science-in-the-making," as a threat to their credibility and authority. Feminist theorists of science have often been at pains to point out the social influences on science and to suggest ways to improve science by making it more inclusive and egalitarian. That scientists and science analysts adhere to very different models of science (outcome and norms vs. process and practice) was pointed out to me by Shirley Strum as we puzzled over the many difficulties that our workshop of scientists and science analysts experienced in attempting to discuss the factors that caused primatologists to change their views of primate society.[80] Often we could not even begin our discussion of the history of ideas in primatology because the participants had such different views of what science is and how it works. We finally realized that we were enacting a local battle in the larger science wars and that C. P. Snow's model of two cultures still holds sway.

To return to the original puzzle at the heart of this paper: Why have so many primatologists incorporated feminist values if some of them so distrust feminism? Why not reject the tenets of feminism along with the label? Many primatologists I have talked to say that they changed their practices in order to make their science better, not necessarily because feminists thought they should do so but because it was right, scientifically right, to flesh out the picture of female primates, to consider questions from a female, as well as a male, perspective, and to research issues of concern to women as well as men. This suggests to me that the goals of feminists and of scientists may sometimes dovetail—at least the goal of producing a better, more inclusive science, one that incorporates the female perspective of both the primatologists and the animals that they study.

Haraway says that several of the women primatologists she interviewed in the 1980s reported affirmation and legitimization for focusing on females in their scientific work from the atmosphere of feminism in their own societies.[81] The men she interviewed also reported legitimization for taking females seriously from the prominence of feminist ideas in their culture and from their friendships with women influenced by feminism. I agree with Haraway that feminism can be credited with creating the atmosphere for legitimizing a new sensitivity to gender issues on the part of primatologists. Thus I think that feminism has been a significant influence on the field of primatology as it exists today, and I have provided many examples of how primatologists have used tools of gender analysis to self-correct in response to the feminist critique of science. However, if feminism and primate science do represent two different cultures, then what we have here may be a case of cultural assimilation, where primatologists have incorporated those feminist values seen to be useful while continuing to reject the "name" and the activism of feminism, which in a scientist's view imply partisan assumptions. Whether primatology will ever move beyond its "two-culture" perspective and its worries about pink ribbons and partisanship and truly become the goddess's discipline remains to be seen.

NOTES

My research is funded by an ongoing grant from the Natural Sciences and Engineering Research Council of Canada (NSERCC). I thank Sandra Zohar for suggestions that improved this paper, and Brian Noble, Naomi Quinn, Londa Schiebinger, Shirley Strum, Zuleyma Tang Martinez, and Alison Wylie for stimulating conversations about gender and science that have shaped my understanding of feminist primatology.

1. Hilary Rose, *Love, Power, and Knowledge: Towards a Feminist Transformation of the Sciences* (Bloomington: Indiana University Press, 1994).

2. Donna Haraway, *Primate Visions: Gender, Race, and Nature in the World of Modern Science* (New York: Routledge, 1989); Sue Rosser, "The Relationship between Women's Studies and Women in Science," in *Feminist Approaches to Science,* ed. Ruth Bleier (New York: Pergamon Press, 1986), 165–80.

3. See the following reviews of Haraway, *Primate Visions:* Susan Cachel, "Partisan Primatology," *American Journal of Primatology* 22 (1990): 139–42; Matt Cartmill, *International Journal of Primatology* 12 (1991): 67–75; Robin Dunbar, "The Apes as We Want to See Them," *New York Times Book Review,* Jan. 10, 1990, 30; Alison Jolly and Margaretta Jolly, "A View from the Other End of the Telescope," *New Scientist,* Apr. 21, 1990, 58; Peter S. Rodman, "Flawed Vision: Deconstruction of Primatology and Primatologists," *Current Anthropology* 31 (1990): 484–86; Meredith F. Small, *American Journal of Physical Anthropology* 82 (1990): 527–28; Craig B. Stanford, *American Anthropologist* 93 (1991): 1031–32.

4. Londa Schiebinger, "Creating Sustainable Science," in *Women, Gender, and Science: New Directions,* ed. Sally G. Kohlstedt and Helen E. Longino, special issue of *Osiris* 12 (1997): 201–16. See also Londa Schiebinger, *Has Feminism Changed Science?* (Cambridge, MA: Harvard University Press, 1999).

5. Haraway, *Primate Visions.*

6. Ibid., 285; Sarah B. Hrdy, introduction to *Female Primates: Studies by Women Primatologists,* ed. Meredith F. Small (New York: Alan R. Liss, 1984), 103–9; and Sarah B. Hrdy, "Empathy, Polyandry, and the Myth of the Coy Female," in *Feminist Approaches to Science,* ed. Ruth Bleier (New York: Pergamon Press, 1986), 119–46.

7. Linda Marie Fedigan, "Is Primatology a Feminist Science?" in *Women in Human Evolution,* ed. Lori Hager (London: Routledge, 1997), 56–75.

8. Schiebinger, *Has Feminism Changed Science?* 8, 186–90.

9. Ibid.

10. Donna Haraway, "Morphing in the Order: Flexible Strategies, Feminist Science Studies, and Primate Revisions," in *Primate Encounters: Models of Science, Gender, and Society,* ed. Shirley C. Strum and Linda Marie Fedigan (Chicago: University of Chicago Press, 2000), 398–420.

11. Linda Marie Fedigan, "The Changing Role of Women in Models of Human Evolution," *Annual Review of Anthropology* 15 (1986): 25–66.

12. Linda Marie Fedigan, *Primate Paradigms: Sex Roles and Social Bonds* (Montreal: Eden Press, 1982; 2d ed., with new introduction, Chicago: University of Chicago Press, 1992).

13. Jane Lancaster, "In Praise of the Achieving Female Monkey," *Psychology Today,* Sept. 1973, 30–36, 99.

14. Altmann in Elisabeth A. Lloyd, "Science and Anti-science: Objectivity and Its Real Enemies," in *Feminism, Science, and the Philosophy of Science*, ed. Lynn H. Nelson and Jack Nelson (Dordrecht: Kluwer Academic Publishers, 1996), 217–59; Hrdy, "Empathy"; Thelma E. Rowell, introduction to section 1, "Mothers, Infants, and Adolescents," in Small, *Female Primates*, 13–16; Meredith F. Small, *Female Choices: Sexual Behavior of Female Primates* (Ithaca, NY: Cornell University Press, 1993); Smuts in Elisabeth Rosenthal, "The Forgotten Female," *Discover*, Dec. 12, 1991, 22–27.

15. Jeanne Altmann, *Baboon Mothers and Infants* (Cambridge, MA: Harvard University Press, 1980).

16. For example, Linda Marie Fedigan, *Primate Paradigms*; David G. Post, "Feeding and Ranging Behavior of the Yellow Baboon" (Ph.D. diss., Yale University, 1978); Katherine Ralls, "Mammals in Which Females Are Larger Than Males," *Quarterly Review of Biology* 51 (1976): 245–76.

17. Barbara B. Smuts, "Male Aggression against Women: An Evolutionary Perspective," *Human Nature* 3 (1992): 1–44; Barbara B. Smuts, "Male Aggression and Sexual Coercion of Females in Nonhuman Primates and Other Mammals: Evidence and Theoretical Implications," in *Advances in the Study of Behavior*, vol. 22, ed. Peter J. Slater, Jay S. Rosenblatt, Charles T. Snowdon, and Manfred Milinski (New York: Academic Press, 1993), 1–63; Barbara B. Smuts, *Sex and Friendship in Baboons* (New York: Aldine, 1985); Barbara B. Smuts, "Gender, Aggression, and Influence," in *Primate Societies*, ed. Barbara B. Smuts, Dorothy L. Cheney, Robert M. Seyfarth, Richard W. Wrangham, and Thomas T. Struhsaker (Chicago: University of Chicago Press, 1987), 400–412.

18. Sarah B. Hrdy, *The Langurs of Abu: Female and Male Strategies of Reproduction* (Cambridge, MA: Harvard University Press, 1977); see also Glenn H. Hausfater and Sarah B. Hrdy, *Infanticide: Comparative and Evolutionary Perspectives* (New York: Aldine, 1984); Sarah B. Hrdy, "Care and Exploitation of Nonhuman Primate Infants by Conspecifics Other than the Mother," in *Advances in the Study of Behavior*, vol. 6, ed. Jay S. Rosenblatt, Robert A. Hinde, Evelyn Shaw, and Colin Beer (New York: Academic Press, 1976), 101–58; Sarah B. Hrdy, *The Woman That Never Evolved* (Cambridge, MA: Harvard University Press, 1981); Sarah B. Hrdy and Debra S. Judge, "Darwin and the Puzzle of Primogeniture: An Essay on Biases in Parental Investment after Death," *Human Nature* 4 (1993): 1–45.

19. Lisa M. Rose and Linda Marie Fedigan, "Vigilance in White-Faced Capuchins, *Cebus capucinus*, in Costa Rica," *Animal Behavior* 49 (1995): 63–70; Lisa Gould, Linda Marie Fedigan, and Lisa M. Rose, "Why Be Vigilant? The Case of the Alpha Male," *International Journal of Primatology* 18 (1997): 401–14; L. Rose, "Costs and Benefits of Resident Males to Females in White-Faced Capuchins," *American Journal of Primatology* 32 (1994): 235–48.

20. Small, *Female Primates*; Small, *Female Choices*; Meredith F. Small, *What's Love Got to Do with It? The Evolution of Human Mating* (New York: Doubleday, 1995).

21. Jane Lancaster, "Carrying and Sharing in Human Evolution," *Human Nature* 1 (1978): 82–89; Jane Lancaster and Chet Lancaster, "Parental Investment: The Hominid Adaptation," in *How Humans Adapt: A Biocultural Odyssey*, ed. Donald J. Ortner (Washington, DC: Smithsonian Institution Press, 1983), 33–65; Jane Lancaster, "The Watershed: Change in Parental Investment and Family Formation Strategies in the Course of Human Evolution," in *Parenting across the Life Span: Biosocial Dimensions*, ed. Jane Lancaster, Jeanne Altmann, Alice Rossi, and Lonnie Sherrod (New York: Aldine, 1987), 187–205; Jane Lancaster, "Women in Biosocial Perspective," in *Gender and Anthropology: Critical Reviews for Research and Teaching*, ed. Sandra Morgen (Washington, DC: American Anthropological Association, 1989), 95–115; Jane Lancaster, "The Evolutionary Bi-

ology of Women," in *Milestones in Human Evolution,* ed. Alan J. Almquist and Anne Manyak (Prospect Heights, IL: Waveland Press, 1993), 21–37.

22. Schiebinger, *Has Feminism Changed Science?* 187.

23. Jeanne Altmann, "Observational Study of Behavior: Sampling Methods," *Behaviour* 49 (1974): 227–67.

24. Lloyd, "Science and Anti-science."

25. Haraway, *Primate Visions,* 304–10; and Lloyd, "Science and Anti-science," 241.

26. For example, Sherwood L. Washburn and C. S. Lancaster, "The Evolution of Hunting," in *Man the Hunter,* ed. Richard B. Lee and Irven Devore (Chicago: Aldine, 1968), 293–303; see discussion in Fedigan, "The Changing Role of Women."

27. Shirley C. Strum and Linda Marie Fedigan, "Theory, Method, and Gender: What Changed Our Views of Primate Society?" in *The New Physical Anthropology,* ed. Shirley C. Strum, Don G. Lindburg, and David A. Hamburg (New York: Prentice Hall, 1999).

28. For example, Clifford J. Jolly, "The Seed-Eaters: A New Model of Hominid Differentiation Based on a Baboon Analogy," *Man* 5 (1970): 5–26; Adrienne Zihlman, "Women in Human Evolution, Part II, Subsistence and Social Organization among Early Hominids," *Signs: Journal of Women in Culture and Society* 4 (1978): 4–20.

29. For example, Karen Strier, "The Myth of the Typical Primate," *Yearbook of Physical Anthropology* 37 (1994): 233–71.

30. Schiebinger, *Has Feminism Changed Science?* 113–18, 189.

31. For example, John H. Crook, "Sexual Selection, Dimorphism, and Social Organization in the Primates," in *Sexual Selection and the Descent of Man, 1871–1971,* ed. Bernard Campbell (Chicago: Aldine Press, 1972), 231–81; Robin Fox, "Alliance and Constraint: Sexual Selection and the Evolution of Human Kinship Systems," in ibid., 282–331.

32. For example, R. D. Martin, "The Biological Basis of Human Behavior," in *The Biology of Brains,* ed. W. B. Broughton (London: Institute of Biology, 1974), 215–50; Hilary Callan, *Ethology and Society: Towards an Anthropological View* (Oxford: Clarendon Press, 1970); Fedigan, *Primate Paradigms.*

33. Haraway, *Primate Visions.*

34. Hrdy, "Empathy."

35. See Linda Marie Fedigan, "Gender Encounters," in Strum and Fedigan, *Primate Encounters,* 498–520.

36. Linda Marie Fedigan, "Science and the Successful Female: Why There Are So Many Women Primatologists," *American Anthropologist* 96 (1994): 529–40; and unpublished data.

37. Schiebinger, *Has Feminism Changed Science?* 188–89.

38. Fedigan, *Primate Paradigms.*

39. Anne I. Dagg, *Harems and Other Horrors: Sexual Bias in Behavioral Biology* (Waterloo, ON: Otter Press, 1983); Katherine Ralls, "Sexual Dimorphism in Mammals: Avian Models and Unanswered Questions," *American Naturalist* 111 (1977): 917–38.

40. Charles Darwin, *The Descent of Man and Selection in Relation to Sex* (London: John Murray, 1871); John H. Crook, "Gelada Baboon Herd Structure and Movement: A Comparative Report," in *Play, Exploration, and Territory in Mammals,* ed. P. A. Jewell and Caroline Loizos, Symposia of the Zoological Society of London, vol. 18 (New York: Academic Press, 1966), 237–58; Hans Kummer, "Social Organization of Hamadryas Baboons: A Field Study," *Bibliotheca Primatologica* 6 (1968): 1–189; cf. Christian Bachmann and Hans Kummer, "Male Assessment of Female Choice in Hamadryas Baboons," *Behavioral Ecology and Sociobiology* 6 (1980): 315–21.

41. Frank A. Beach, "Sexual Attractivity, Proceptivity, and Receptivity in Female Mammals," *Hormones and Behavior* 7 (1976): 105–38.

42. Donna J. Haraway, *Simians, Cyborgs, and Women* (New York: Routledge, 1991).

43. Darwin, *The Descent of Man*.

44. Fedigan, *Primate Paradigms*.

45. For example, John D. Berard, Peter Nurnberg, Jorg T. Epplen, and Jorg Schmidtke, "Male Rank, Reproductive Behavior, and Reproductive Success in Free-Ranging Rhesus Monkeys," *Primates* 34 (1993): 481–89; Pascal Gagneux, David S. Woodruff, and Christophe Boesch, "Furtive Mating in Female Chimpanzees," *Nature* 387 (1997): 358–59; Hideyuki Ohsawa, Miho Inoue, and Osamu Takenaka, "Mating Strategy and Reproductive Success of Male Patas Monkeys (*Erythrocebus patas*)," *Primates* 34 (1993): 533–44; Yukimaru Sugiyama, Sakie Kawamoto, Osamu Takenaka, Kiyonori Kumazaki, and Norikatsu Miwa, "Paternity Discrimination and Inter-group Relationships of Chimpanzees at Bossou," *Primates* 34 (1993): 545–52; Richard W. Wrangham, "Subtle, Secret Female Chimpanzees," *Science* 277 (1997): 774–75.

46. Schiebinger, *Has Feminism Changed Science?* 189.

47. Catherine Lutz, "The Gender of Theory," in *Women Writing Culture*, ed. Ruth Behar and Deborah A. Gordon (Berkeley and Los Angeles: University of California Press, 1995), 249–66.

48. Bruno Latour, "A Well-Articulated Primatology: Reflexions of a Fellow-Traveler," in Strum and Fedigan, *Primate Encounters*, 358–81.

49. Haraway, *Primate Visions*.

50. See Linda Marie Fedigan and Laurence Fedigan, "Gender and the Study of Primates," in *Gender and Anthropology: Critical Reviews for Teaching and Research*, ed. Sandra Morgen (Washington, DC: American Anthropological Association, 1989), 41–64.

51. Altmann, *Baboon Mothers and Infants*.

52. Thelma E. Rowell, "The Concept of Dominance," *Behavioral Biology* 11 (1974): 131–54.

53. Shirley C. Strum, *Almost Human: A Journey into the World of Baboons* (New York: Random House, 1987).

54. Smuts, *Sex and Friendship in Baboons*.

55. Richard W. Wrangham, "An Ecological Model of Female-Bonded Primate Groups," *Behaviour* 75 (1980): 262–300.

56. For example, Donald S. Sade, "Some Aspects of Parent-Offspring and Sibling Relations in a Group of Rhesus Monkeys, with a Discussion of Grooming," *American Journal of Physical Anthropology* 23 (1965): 1–18; Donald S. Sade, "Inhibition of Mother-Son Mating among Free-Ranging Rhesus Monkeys," *Science and Psychoanalysis* 12 (1968): 18–38; Donald S. Sade, "A Longitudinal Study of Social Relations of Rhesus Monkeys," in *Functional and Evolutionary Biology of Primates*, ed. Russell H. Tuttle (Chicago: Aldine, 1972), 378–98.

57. For example, Hrdy, *The Woman That Never Evolved*.

58. Hrdy, "Care and Exploitation"; Hrdy, *The Langurs of Abu*; and Hrdy, "Empathy."

59. For example, Haraway, *Primate Visions*.

60. Michael A. Huffman, "Mate Selection and Partner Preferences in Female Japanese Macaques," in *The Monkeys of Arashiyama: Thirty-five Years of Research in Japan and the West*, ed. Linda Marie Fedigan and Pamela J. Asquith (New York: State University of New York Press, 1991), 101–22; Charles H. Janson, "Female Choice and Mating System of the Brown Capuchin Monkey *Cebus apella* (Primates: Cebidae)," *Zeitschrift für Tierpsychologie* 65 (1984): 177–200; Joseph H. Manson, "Measuring Female Mate Choice in Cayo Santiago Rhesus Macaques," *Animal Behavior* 44 (1992): 405–16; Small, *Female Choices*.

61. Karen B. Strier, "Brotherhoods among Atelines: Kinship, Affiliation, and Competition," *Behaviour* 130 (1994): 151–67.

62. For example, Karen B. Strier, "The Myth of the Typical Primate," *Yearbook of Physical Anthropology* 37 (1994): 233–71.

63. Sue Boinski, "Vocal Coordination of Troop Movement in Squirrel Monkeys (*Saimiri oerstedi* and *S. sciureus*) and White-Faced Capuchins (*Cebus capucinus*)," in *Adaptive Radiations of Neotropical Primates,* ed. Marilyn A. Norconk, Alfred L. Rosenberger, and Paul A. Garber (New York: Plenum Press, 1996), 251–69.

64. Carolyn M. Crockett, "Emigration by Female Red Howler Monkeys and the Case for Female Competition," in Small, *Female Primates,* 159–73; Kenneth E. Glander, "Dispersal Patterns in Costa Rican Mantled Howling Monkeys," *International Journal of Primatology* 13 (1992): 425–36; Margaret R. Clarke and Kenneth E. Glander, "Female Reproductive Success in a Group of Free-Ranging Howling Monkeys (*Alouatta palliata*) in Costa Rica," in Small, *Female Primates,* 111–26.

65. Devra G. Kleiman, "Monogamy in Mammals," *Quarterly Review of Biology* 52 (1977): 39–69; Patricia C. Wright, "Biparental Care in *Aotus trivirgatus* and *Callicebus moloch,*" in Small, *Female Primates,* 59–75.

66. Alison Jolly, "The Puzzle of Female Feeding Priority," in Small, *Female Primates,* 197–215; Michael E. Pereira, Ruben Kaufman, Peter M. Kappeler, and Deborah J. Overdorff, "Female Dominance Does Not Characterize All of the Lemuridae," *Folia primatologica* 55 (1990): 96–103; Alison Richard, "Malagasy Prosimians: Female Dominance," in Smuts et al., *Primate Societies,* 25–33.

67. Schiebinger, *Has Feminism Changed Science?* 190.

68. Sarah B. Hrdy, introduction to Small, *Female Primates,* 103–9; Hrdy, "Empathy"; Rowell, introduction to section 1 in Small, *Female Primates;* P. Sicotte and C. Nisan, "Femmes et empathie en primatologie," in *Grands singes: la fascination du double,* ed. Bertrand L. Deputte and Jacques Vauclair (Paris: Autrement, 1998), 77–102; Small, review of *Primate Visions* in *American Journal of Physical Anthropology* 82 (1990): 527–28; Strum, *Almost Human,* and unpublished manuscript.

69. George B. Schaller, *The Year of the Gorilla* (Chicago: University of Chicago Press, 1964).

70. See Pamela J. Asquith, "Primate Research Groups in Japan: Orientations and East-West Differences," in Fedigan and Asquith, *The Monkeys of Arashiyama,* 81–98; H. Takasaki, "Traditions of the Kyoto School of Field Primatology in Japan," in Strum and Fedigan, *Primate Encounters,* 151–64.

71. See R. A. Mittermeier, "A Global Overview of Primate Conservation," in *Primate Ecology and Conservation,* ed. James G. Else and Phyllis C. Lee (Cambridge: Cambridge University Press, 1986), 325–40.

72. For example, Harold T. P. Hayes, *The Dark Romance of Dian Fossey* (New York: Simon and Schuster, 1990); Marguerite Holloway, "Profile: Jane Goodall—Gombe's Famous Primate," *Scientific American,* Oct. 1997, 42–44; Sy Montgomery, *Walking with the Great Apes* (Boston: Houghton Mifflin, 1991); Virginia Morell, "Called 'Trimates,' Three Bold Women Shaped Their Field," *Science* 260 (1993): 420–25; Brian Noble, "Politics, Gender, and Worldly Primatology: The Goodall-Fossey Nexus," in Strum and Fedigan, *Primate Encounters,* 436–62; Linda Spalding, *The Follow* (Toronto, ON: Key Porter Press, 1998).

73. Haraway, *Primate Visions,* 182.

74. Alison Wylie, "Standpoint Matters, in Archaeology for Example," in Strum and Fedigan, *Primate Encounters,* 243–60.

75. Fedigan, "Science and the Successful Female"; Haraway, *Primate Visions;* Hrdy, "Empathy."

76. Margaret W. Rossiter, "Which Science? Which Women?" *Osiris* 12 (1977): 169–85.

77. Schiebinger, *Has Feminism Changed Science?*

78. C. P. Snow, *Two Cultures and the Scientific Revolution* (New York: American Library, 1959).

79. Steven Shapin, *The Social History of Truth* (Chicago: University of Chicago Press, 1994); Bruno Latour, *Science in Action: How to Follow Scientists and Engineers through Society* (Cambridge, MA: Harvard University Press, 1987); and Latour, "A Well-Articulated Primatology."

80. Shirley C. Strum, "Science Encounters," in Strum and Fedigan, *Primate Encounters*, 475–97.

81. Haraway, "Morphing in the Order."

FOUR

Revisiting Women, Gender, and Feminism in Developmental Biology

SCOTT F. GILBERT AND KAREN A. RADER

Figure 4.1 shows eight biologists—three women and five men—sitting together at a lecture at the annual meeting of the Society for Developmental Biology in 1998. This is an interesting photograph if only because it shows a fairly equal representation of senior men and women at a scientific conference. But this picture could also represent a poor sample, a subset too small to represent the entire group. Indeed, it is: it is a picture only of the recent presidents of the society. ·

Women appear to have done extremely well in developmental biology, both in scientific research and in ascending its professional ranks. As Evelyn Fox Keller has noted, "it is the intellectual space occupied by women in developmental biology today that has led to the subjective impression among some biologists that developmental biology is a field now dominated by women."[1] Most prominently, the first Nobel Prize awarded to developmental biologists in fifty years went in 1995 to Christiane Nüsslein-Volhard (who won the prize along with her colleague Eric Wieschaus and the geneticist Edward B. Lewis), and the first March of Dimes Award in Developmental Biology went jointly to Beatrice Mintz and Ralph Brinster in 1996. Of the fourteen members of the present executive board of the Society of Developmental Biology, nine are women, including its president and seven of the nine members-at-large. Any discussion of who are the most influential developmental biologists in the world would include (but certainly not be limited to) such names as Kathryn Anderson, Cori Bargmann, Ruth Bellairs, Marianne Bronner-Fraser, Connie Cepko, Marie Di Berardino, Elizabeth Hay, Brigid Hogan, Vivian Irish, Laurinda Jaffe, Cynthia Kenyon, Judith Kimble, Nicole Le Douarin, Ruth Lehmann, Gail Martin, Anne McLaren, Barbara Meyer, Lee Niswander, Virginia

Figure 4.1 Former presidents of the Society for Developmental Biology assembled during the Conklin Lecture at the 1998 annual society meeting: (*top, left to right*) Dave McClay, Matt Scott, Chuck Kimmel, Helen Blau, Janet Rossant, (*bottom*) Kathryn Anderson (appearing only partially), Alan Spradling, Meredith Runner. Photograph taken by Laurie Iten for the society's website.

Papaiannou, Liz Robertson, Janet Rossant, Carla Schatz, Trudy Schüpbach, Irma Thesleff, Cheryll Tickle, Shirley Tilghman, Kathryn Tosney, and Virginia Walbot.

Like any important and anomalous observation in science, the apparent success of women in developmental biology suggests more questions than it answers. For though the number of women who have recently received assistant professorships in this field is remarkable, the total number of women practitioners is still under 50 percent, as Keller also notes. Thus the most basic questions are: What constituted the success of women developmental biologists and how did it come about? In the era from 1930 to the present, when feminists have been increasingly concerned about professional gains made by women in science, how did developmental biology attract and support a relatively large number and variety of women? How did particular individuals negotiate careers as developmental biologists in ways that allowed them to be perceived as leaders in this field from its start, and did these strategies and perceptions change over time? Another, more complicated question follows from this line of inquiry: namely, how have the number and achievements of women in developmental biology during this period made a difference? Have these women made developmental biology a "feminist science"—or has feminism changed the means by which we do developmental biology in other ways?

In her essay "Developmental Biology as a Feminist Cause?" Keller addresses many of these issues. Keller suggests that the large number of women in developmental biology "has a lot to do with timing."[2] For the period since World War II, she cites the coincidence of increasing numbers of women in science and the rise of developmental biology as a field (though the disciplinary label itself only dates from the 1960s).[3] For the earlier period, she notes that the type of scientific work developmental biology required "was hard, often back-breaking work and widely assumed to be unrewarding. What more natural job to assign to women?"[4] The first goal of our chapter is to further contextualize the history of women developmental biologists in relation to specific practical and institutional circumstances in biology—both before and after World War II—and suggest some additional areas for exploration.

We also want to revisit the issue of the meaning of gender in the history of developmental biology in order to inquire how it might be investigated further. Along these lines, Keller argues that the career of Nüsslein-Volhard illustrates the potency of the cultural symbolic work of gender in the history of developmental biology. Nüsslein-Volhard, she demonstrates, possessed a "multifaceted ambivalence" about feminism and the transformation of scientific career tracks in order to accommodate or encourage women. But it was precisely her ambivalence that situated her to make an "intervention of immense value to women in science"—specifically, as a mentor to some American women developmental biologists and as a researcher who sought to restore investigative prominence to the role played by the egg's cytoplasm in gene activation. "Nüsslein-Volhard," Keller writes, "stood at the intersection of multiple crossroads, able to make remarkably productive use of the ambiguities of her location in large part because of the timing of her intervention."[5] Using our own brief case studies of Salome Waelsch and C. H. Waddington, we argue that Keller's emphasis on the power of multiple "situatedness" for women developmental biologists might be broadly generalizable to early practitioners in the field as a whole. Thus we suggest that gender would be a potent historical tool for exploring the social and intellectual history of developmental biology as it relates to the broader history of twentieth-century biology, as well as to the lives and work of individual scientists.

Ultimately, we discuss the historical intersection of late-twentieth-century feminism with developmental biology and point to how the knowledge critiques that resulted transformed the field. But just as there is no one feminism, there is no one feminist critique of science, nor is there any one reason for any particular woman to enter science or any field of science. What attracts one woman to a science may repel another. A feminist scientific agenda of one age might be the reactionary agenda of a different age. We conclude, then, that while these critiques are the best places to look for the difference that feminism has made thus far in developmental biology, much historical and sociological work remains to be done on the fate of feminist ideals in both the theory and the practice of this growing scientific discipline.

WOMEN, EMBRYOLOGY, AND GENDER BEFORE WORLD WAR II: A DYNAMIC OF INSTITUTIONAL AND SOCIAL RESOURCES

We should look first at the issue of how women first came to occupy the field of developmental biology.[6] What historical conditions might have allowed women to find this particular niche in the sciences? That is, how would women be informed that there even was such a field as developmental biology?

The answer to this question may change dramatically with the politics of the times, but in early-twentieth-century America, there were no obvious intellectual incentives in the standard public school curriculum for girls or women to learn about research in embryology/developmental biology. Developmental biology is not a subject that has ever been well integrated in high school biology books. Indeed, probably very few of us were taught developmental biology in our high schools, because to teach developmental biology means teaching sex, and we cannot do that in America. Contemporary developmental biology is a niche more likely to be presented in media than in textbooks and talked about more in schoolyards than in classrooms. High school biology books are characterized by gorgeous pictures and superficial discussion. The Biological Science Curriculum Study (BSCS), which has published some of the most important high school biology textbooks in the past thirty years, set up its first developmental biology advising group as late as 1999. The title of the vanguard BSCS book, though, is no longer called *From Molecules to Man*.

One explanation that needs to be empirically investigated is whether the social and material situatedness of women's bodies in any way contributed to women's entering this field. Development from the human zygote to the newborn human being is a process that takes place within the body of a woman and that never happens within the body of a man. To the extent that having a vagina, ovaries, and a womb has been ideologically important in Anglo-American culture, and to the extent that being fertile has been considered important to the family and the nation, embryology could have been construed most literally as "women's work." And if one were already training in biology and looking for a field in which to specialize, one might ask, "Are the questions of this field fundamental and important questions?" Because of her specific cultural location, a middle-class woman coming of age in the early twentieth century might perceive the questions of embryology to be important and worthy of further investigation—much in the same way that many women were drawn to eugenics research "by sympathy with its ideals."[7]

But once there, what conditions might have allowed women to find this field more comfortable than other possible scientific fields? Here other obvious reasons emerge which concern the unique institutional configuration of embryology and its corresponding place in the professional hierarchy. At the turn of the last century, teaching was considered a role where women could influence the world; it certainly gave women public responsibility and got them out of the home. As Margaret Rossiter has pointed out, natural history and its teaching became open to women in the 1870s, and embryology was seen as being an excellent and accessible entry into the world of nature.[8] The opening of a chick's egg each day during its three-week incubation provides a wonderful view of development, as does the metamorphosis of tadpoles and caterpillars. Embryology has claimed a large number of women practitioners since its inception in America, and this seems to be intimately connected with biology education. The Marine Biological Laboratories (MBL) at Woods Hole was founded by collaboration between the Boston Society of Natural History and the Women's Education Association of Boston.[9] The embryology courses at the MBL were evenly filled by men and women (although the instructors were routinely male).

But though embryology was initially considered to be one of the most important elements of natural history,[10] women began to be excluded from this and other sciences when the urge to professionalize swept academia in the 1890s. The MBL was no exception to this trend, even though the women scientists there were already well established. The women from Goucher, Mount Holyoke, and Bryn Mawr would still come to the MBL, but they did not get positions in the prestigious universities; instead, they brought natural history into high schools and women's colleges. Not insignificantly, they also brought their expertise into their husband's laboratories. E. B. Wilson, T. H. Morgan, E. Conklin, F. R. Lillie, and E. N. Harvey each found his wife-to-be at the MBL. Conklin wrote that marriages might be made in heaven, "but there is certainly a large branch office in Woods Hole."[11]

Relatedly, in the 1920s, classical genetics displaced embryology from its position of being the major biological science explaining heredity, and this newer, more reductionist discipline was almost entirely male at its cutting edge. Looking at T. H. Morgan's laboratory, which was to become the paradigm for genetic research centers, Robert Kohler notes, "Wives of graduate students worked as technicians and stockkeepers. So the village society of the drosophilists was not monkishly male, but women did not occupy official positions; they were there as unpaid working wives and volunteers. They do not appear in official photographs. The group's formative psychosocial relationships were male: master and disciple, father and son, Boss and the 'boys.' "[12] Morgan did have some women graduate students, but they were placed on peripheral projects (not the gene-mapping one) and published fewer papers than the "boys." Thus in the first three issues of *Genetics* (starting in 1916), there are no women authors. The sole woman author in volume 4 is Clara Lynch, a doctoral student of Morgan's who was doing her thesis on interspecific sterility and who later left drosophila genetics to pursue work at Rockefeller University on what Kohler has called the "messier aspects" of genetic problems in mice. But even those who began wanting to work on "messy organisms" did not fare much better. As late as 1928, the president of Harvard rejected the application of a Miss Warmbier to the Bussey Institution—Harvard's preeminent mammalian and plant genetics research center—on the grounds that her place might be more productively filled by a male student.[13] In short, genetics research was at the forefront of American life sciences both intellectually and professionally, and with the prominent exception of eugenics fieldwork, women were difficult to find.[14]

In turn, embryology was marginalized and lost its former prestige.[15] Until 1995, only one embryologist (Hans Spemann) had received a Nobel Prize. In many ways, it may be comparable to X-ray crystallography, another field that was considered peripheral, full of material details, and full of women practitioners. With genetics attracting the men (who, after all, were considered the employable members of society), embryology was left to women, who could get positions at teaching colleges, women's colleges, and private foundations or research institutions.

It would also be interesting to determine if the material culture of embryology further contributed to women's professional advancement in ways that other life science practice could not. For example, dependence on animal breeding seasons presents potential pushes and pulls for women entering this field. Until the current age of molecular techniques—that is, before today's professors got their positions—embryology was not an easy subject in which to make a reputation. If you wanted to study amphibian development, you waited until spring, went out into the woods, collected the freshly laid eggs, and did your experiments as fast as you could. Then you had all summer, winter, and fall to fix, section, stain, and analyze your data. For example, Hilde Mangold's work on "the organizer" in Hans Spemann's laboratory took two breeding seasons to finish. The first group of experiments did not give definitive results, and she had to wait until the next spring's rain brought new clutches of eggs. This slower timetable may have been advantageous from the perspective of women who wanted both to do science and to raise children: one could more easily become as good an embryologist as any man and still tend to one's family. But also, as C. H. Waddington noted, other biological sciences (especially genetics) gave results much faster.[16] Since (then, as now) the number of publications counted toward tenure and promotion, men might see embryology as a difficult way to earn a living, and therefore, women might have more readily found viable careers doing this kind of biological work.[17]

Furthermore, as anybody who has worked with embryos knows, embryology, especially as it existed until the age of molecular techniques, demands fine motor skills. Manual dexterity was not just important—it was essential. One had to love precise and detailed movements with needles. One teased out pieces of somites, regions of notochords, even individual cells with one's needles and one's fingers. Because women of the time were encouraged to master needlework and other such crafts, these practical factors may initially have been significant for encouraging some women to enter a scientific field that required the same skills.

ENTRY AND SUCCESS IN DEVELOPMENTAL BIOLOGY: SALOME GLUECKSOHN WAELSCH AND C. H. WADDINGTON

Exploring the areas we have described thus far would give us even more historical information about how questions of gender related to the early involvement of women in developmental biology—specifically, embryology. But another question about women's participation remains: though many newly trained women scientists in the period from 1900 to 1940 pursued embryologically oriented fields, were they uniquely flourishing there—and why or why not? Examining in more detail the early careers of two developmental biologists—Salome Gluecksohn Waelsch (1907–; hereafter referred to as Waelsch, although she published under various names) and C. H. Waddington (1906–75)—is instructive for understanding the professional world faced by early-twentieth-century developmental biologists, both men and women, and how gender shaped the way particular individuals negotiated places for themselves in this world.

By her own admission, Waelsch "wasn't planning to be a scientist" when she began university training. In school, though Waelsch was a very good student and had at least one woman teacher she "really respected and loved," she also had to endure the persistent anti-woman and anti-Semitic taunts of her classmates. College in Konigsberg was a welcome relief, and she originally intended to become a humanist: a Classics teacher. But like other women developmental biologists of her generation, Waelsch first came to her career in science, not because of an innate passion for the subject, but because she thought it would be the most practical route to a desired career in teaching. Once Waelsch decided to study biology, a combination of fate and persistence led her to doctoral studies. In order to earn a living to supplement her scholarships, she became a tutor to a family in Berlin. The family asked Waelsch if she would consider moving with her charge to a smaller town: "I was asked to choose a town. I chose Freiburg, because by that time I had become interested in developmental biology."[18] The University of Freiburg was the home of Hans Spemann, an already distinguished experimental embryologist and soon (1935) to be Nobel laureate for his work with "the organizer."

Not unlike other women who entered graduate programs in biology about this time, Waelsch characterizes her first experience with the world of professional academic science as "stimulating to the utmost"[19] but "negative in essence."[20] Spemann proved a reluctant teacher and an impossible mentor. Though Waelsch thought Spemann's embryology was "very exciting"[21]—as compared to genetics, which "was not my thing"[22]—she found him to be "old and an anti-Semite, and also a strong anti-feminist to participants in his experiments. He was not very eager to take me in."[23] In practice, this meant that although Spemann accepted her as a student, he assigned her "a rather boring descriptive study of limb development" which he hoped would provide the basis for some exciting experimental work on the roles of ectoderm and mesoderm in neural patterning. The important projects, Waelsch remembers, were assigned to "a young man who became an object of Spemann's love at first sight and who remained his favorite pupil."[24] Interestingly, Waelsch was not Spemann's first or only female graduate student. To this day she frequently comments that some of the most important work to come out of the Freiburg laboratory was completed by similarly mistreated women scientists: specifically, Hilde Mangold's work on the organizer and Else Wehmeier's experiments on embryonic induction.[25]

Although she felt that Spemann discriminated against her intellectually because she was a woman, gender did not determine the division of labor in the laboratory's practical work. For example, during amphibian breeding season (three to four months in early spring), Waelsch made it her "ambition that there would not be a minute in the twenty-four hours of the day that I would not have spent in the lab at some point."[26] But she distinctly remembers these efforts were collaborative among herself and her male junior colleagues: "all of us worked day and night and we shared results, interpretations, etc."[27]

Also, and perhaps more significantly, when it came to critically evaluating the ideas and methods of their senior mentor, nearly all the students—male and female—expressed a belief that Spemann's work was too narrow. Viktor Hamburger, Spemann's senior graduate student and Privatdozent and Waelsch's de facto supervisor, formally arranged joint seminars with the Department of Philosophy to counteract this narrow thinking, and he made sure that the students obtained some introduction to the principles of genetics and how they relate to embryology.[28] Along these lines, Waelsch relied on several supportive male scientific colleagues who proved powerful intellectual and personal resources in her Freiburg years. In addition to her laboratory colleagues Hamburger and Oscar Schotte, she formed a close personal and professional friendship with British biologist C. H. Waddington, who came as a visitor to Spemann's laboratory in 1931. This was the year during which Waelsch had begun to mistrust her mentor's vitalist explanations and to have her own "thoughts about the role of genes and their possible activation in the [developmental] induction mechanism."[29] From Waddington, Waelsch says, she "received much encouragement and infinite stimulation in thinking about problems of development . . . Waddington remained one of my closest friends until the time of his death."[30]

Waddington's own entrée into developmental biology was very different from Waelsch's, although equally circuitous. He graduated from Sidney Sussex College, Cambridge, in 1926 with a degree in geology and began a Ph.D. thesis in paleontology. As Edward Yoxen has noted, this represented "a very classical and academic retreat from the scientific service of an expanding international industry."[31] But Waddington had supreme self-confidence and a decidedly philosophical bent gleaned from years as a member of the progressive Cambridge-based Biotheoretical Gathering (whose regular participants included such distinguished scientists as Gregory Bateson, Evelyn Hutchinson, and Joseph Needham). As a result, unlike Waelsch, he was perceived as ambitious and even as something of a *Wunderkind.* He identified himself as a student of "diachronic biology"—"embryology-genetics-evolution which again form a group whose interconnections are obvious and unavoidable"[32]—though he never obtained his Ph.D. Instead he migrated from work on chick embryo culture (with Dame Honor Fell) at Strangeways Research Laboratory in England to work on amphibian neural induction at Otto Mangold's laboratory in Berlin.

But by the time Waddington came to Spemann's laboratory, his wandering had paid off. He had successfully presented his chick embryo work at the International Congress of Experimental Cytology in Amsterdam, and he was actively seeking research areas in which he could combine his traditional embryological expertise with new molecular and genetic approaches. He continued the chick project in collaboration with biochemists Joseph and Dorothy Needham, in the hopes of identifying the active agent of embryonic induction, and along these lines, he attracted the attention (and support) of the Rockefeller Foundation. But by 1938, this collaborative effort had stalled and Waddington decided to travel to the United States to visit several genetic and developmental research groups. One of the first of these was L. C. Dunn's mammalian genetics group at Columbia University, where Waddington renewed his acquaintance with Waelsch and her work.

Waelsch's developmental work had by this time taken a decidedly genetic turn—in part because of her scientific interests in this conjunction and in part because of contingent historical circumstances. The completion of Waelsch's Ph.D., in 1932, coincided with Hitler's rise to power in Germany and—as for many German scientists and intellectuals—this proved a turning point in her career. In early 1932 she started to look for German postdoctoral positions where she could pursue her interest in the border between genetics and embryology, but she met with resistance. In Richard Goldschmidt's laboratory in Berlin, for example, Waelsch was flatly turned away by Curt Stern (Goldschmidt's assistant), who told her, "You, a woman and a Jew—forget it!"[33] In 1933 she met and married Rudolf Schoenheimer, one of Germany's most promising young biochemists. Schoenheimer strongly supported Waelsch's scientific career, but in private they agreed "that it would be extremely difficult. . . . [O]ur ambitions [to become a dual career couple] were not terribly high."[34] Thus when he was offered a position at Columbia University's College of Physicians and Surgeons, the pair immediately left Berlin for New York. In August 1933, after having spent six months "in a corner . . . without a job or a desk" in Columbia embryologist Robert Detwiler's laboratory, she met Columbia mouse geneticist L. C. Dunn at a dinner party. Waelsch recalls: "He was interested in my experiences and my training and he invited me to come to his laboratory, though he said he couldn't pay me. He had no money."[35] She saw it as a good intellectual opportunity to learn genetics.[36] It would be three years before she was officially appointed as research associate at Columbia—at an annual salary of $1,500. In the meantime, she set about retraining herself with a new complex mammalian system: the mouse.

Waelsch recalls that the atmosphere at Columbia was a far cry from the one she had experienced in Spemann's lab. To begin with, "Dunny" (as Waelsch eventually learned to call him) was politically committed and "progressive," despite the fact that he also "never met a pretty girl that he didn't pursue."[37] Dunn was a leader in the attempts to rescue German Jewish biologists and find positions for them in the United States. Also, unlike in Freiburg, the work itself was not communal: "I learned it [genetics] really by working with the animals. . . . You see, there really was no group, you know? Dobzhansky was in one corner, way back, and Dunn was in the other corner. There was very little contact."[38] Her makeshift office was located between the mouse room and Dunn's office, and each day's contact with the mammals, though "intriguing," brought new technical challenges.[39] Interestingly, Waelsch felt that neither her biology nor her socialization had especially prepared her for these challenges, but this did not limit her scientifically: "I was never particularly good with my hands, but I was perfectly able to do whatever was needed."[40] Waelsch and Dunn's subsequent collaboration consisted primarily of work on so-called T-mutation mice (a dominant mutation called *Brachyury* wherein the heterozygote $T/+$ mice had short tails, and the homozygous mutants died in utero), and their experiments are now hailed as the beginning of developmental genetics.[41]

Waelsch credits Dunn for the foresight to exploit the T-locus as a model system for genetic study,[42] but she was attracted to T-locus work for a different reason: because it showed "numerous and unorthodox aspects of genetic behavior" and was "unwilling to conform to the expectations of conventional genetics." For Waelsch, the real beauty of the T-locus was that it embodied a complex interaction of the structures and processes that were central to many important biological questions, including development: "[In the] T-complex[,] . . . relevant genes were shown to affect a variety of systems, thus creating a diver-

sity of problems including those of genetic transmission, recombina-
tion, gene action, pleiotropy, evolution, genetic control of development, and spermatogenesis. Such a complexity of effects presented a
unique situation as well as opportunity, and raised questions of gene
structure, organization, and expression, many of which have remained
unanswered to this day."[43]

Moreover, the data from the observations on mutant mice fit well
into the organizer project which Spemann had forbidden her to pursue
in Freiburg.[44] Indeed, from her perspective, Waelsch rejects the explanation that her interests were marginal to the mainstream of experimental biology.[45] She felt that others—both male and female—shared
her interests and her aesthetics: "during the middle and late 1930s . . .
I witnessed the expression of a strong liaison between embryology and
genetics . . . and I believe that it may be due in large part to my own
close contacts with particular people."[46]

One of these people was Waddington, who visited Dunn's laboratory and was so convinced of the importance of Waelsch's mouse studies that in 1939 he, too, sought to combine genetics and development
through a collaborative project on *Drosophila* wing deformities with
T. H. Morgan's Caltech genetics research group. What initially motivated this work was Waddington's desire to demonstrate that the embryologist and the geneticist were studying the same phenomena: "In
the late thirties I began developing the notion that the process of becoming (say) a nerve cell should be regarded as the result of a large
number of genes which interact to form a unified 'concrescence.' "[47]
Though many embryologists of the 1930s were wary of what Ross
Harrison deemed geneticists' *Wanderlust* for developmental problems,
Waddington forged ahead, and between 1938 and 1940 he wrote
two textbooks and two review articles concerning the developmental
action of genes. This was quite a presumptive undertaking for a thirty-
three-year-old geologically trained embryologist who had yet to publish his first paper in genetics. But while perhaps full of bravado,
Waddington's vision was strikingly similar to Waelsch's: he sought
to identify neither the inducer nor the mechanism of gene action but
"the whole complex system of actions and interactions which constitute differentiation."[48]

These overlapping biographical narratives highlight both similarities and differences in the early-twentieth-century experience of men
and women practitioners of developmental biology. Clearly, though
Waelsch's and Waddington's respective training was very different,
their experiences led them to a common vision of developmental biology as a discipline that embraced both embryological and genetic prac-

tices. But professionally or practically, one would be hard pressed to argue that Waelsch in any way benefited from being a woman, an outsider, or a person with what we would today call interdisciplinary interests. She did not conceive of her project or her skills as "woman's work," but Columbia clearly did, and predictable institutional mechanisms consequently ensured that her work would be perceived as marginal. It was not until 1953, nineteen years after she began her T-locus work, that she finally obtained an independent appointment there and even then it was not in genetics but as a research associate in the Department of Obstetrics and Gynecology, at the College of Physicians and Surgeons. By contrast, in 1944, only six years after his first publications in genetics, Waddington's boundary crossing was rewarded: he was chosen to head up England's National Animal Breeding and Genetics Research Organization. From here, "he set out what he thought were the important strategic questions in biology." Not everyone listened, but at least he had a sanctioned forum.[49]

At the same time, this exercise suggests that gender might be used to understand the social and intellectual history of developmental biology more broadly—namely, to investigate the disciplinary boundary between developmental biology and genetics. Keller suggests that we pay attention to the cultural symbolic work of gender, and here we find Harrison's *Wanderlust* rhetoric particularly instructive. Regardless of how the practitioners themselves thought of their work, our historical understandings of their experiences would clearly benefit from a more systematic analysis of the rhetorical coding of genetics as "male" and embryology as "female" during this early period.[50]

One particularly fruitful avenue in this regard might be analysis of aesthetics. The emphasis on complexity over simplicity is a traditional characteristic of embryology that separates it from the aesthetics of genetics. The relationship between the aesthetic dimensions of embryology and feminism are briefly explored by Gilbert and Faber.[51] Whether scientists enter certain areas because of aesthetic considerations is a relatively unexplored question, but because aesthetics is infused with gender, it may contribute significantly to the recruitment and sustaining of either men or women in particular areas of science.[52] With such a rubric, we might more carefully generalize about how the kind of ambivalences and boundary crossing that characterize the careers of people like Waelsch, Waddington, and Nüsslein-Volhard map onto the kind of ambivalences embodied by biology itself during this critical period.[53] In other words, we would further illuminate the relationship between the problem of "men and women in biology" and "gender and biology."

WOMEN IN CONTEMPORARY DEVELOPMENTAL BIOLOGY: OPPORTUNITY, COMMUNITY, AND THE FEMINIST CRITIQUE

Meanwhile, however, Keller's injunction points to a more obvious place to explore how gender has shaped developmental biology and the experiences of its practitioners: specifically, the historical concordance of the postwar influx of women into developmental biology with feminist critiques of the field's mainstream scientific ideas. In the 1960s, the women's liberation movement opened up new possibilities for women in the professions. Medical schools, law schools, and even science departments began to accept women into their programs and to hire women as full-time faculty members. Universities that were slow to make the change found that talented women were being drawn away. Women who were interested in the sciences could attend the prestigious schools, and the politics had changed. Rather than become schoolteachers, these women were told (by Marlo Thomas if no one else) that they could be anything they wanted to be—even full professors and PIs.

As Waelsch's career illustrates, those who were already in developmental fields benefited from these developments. Bryn Mawr biologist Jane Oppenheimer, Waelsch's good friend and one of the few women in her cohort to have an official faculty position (albeit at a women's liberal arts college), called Waelsch's work to the attention of anatomist Ernst Scharrer. Scharrer had left his position at the University of Denver to organize his own department at the newly created Albert Einstein College of Medicine in New York: "he saw a chance there to do away with academic prejudices, e.g. against women on faculties of universities and medical schools."[54] Waelsch was among three of Scharrer's first appointments—all of whom were women. Within three years, she was promoted to full professor, and in 1963 she became the first chair of the newly separate Department of Genetics.

Beyond general trends regarding more women participating in science (and the paid professional workforce more generally), the presence of women like Waelsch in prominent places likely attracted more women to this particular type of biological work—a phenomenon Keller calls the "Jewish violinist from Odessa effect."[55] And Waelsch was hardly the only woman. By the mid-1960s, the ranks of women developmental biologists included scientists such as Ruth Bellairs, Anna Ginsburg, Anne McLaren, Kirstie Lawson, Nicole Le Douarin, Hephizibah Eyal-Giladi, and Mary Rawles. Anecdotal evidence suggests that the prominence of these women in their respective areas of developmental biology was influential in attracting younger women into these areas. Indeed, by 1963, half the papers in the *Journal of Experimental Embryology and Morphology* were published by women. It was relatively easy in developmental biology to be trained by a woman (or by a man who had been trained by a woman); there may not be other areas of biology where this can be said. (For instance, SFG was trained as a doctoral student in mammalian developmental genetics by Barbara Migeon; his postdoctoral advisor in mammalian developmental biology was Robert Auerbach, a feminist and former student of Salome Gluecksohn Waelsch.)

To the extent that women scientists wanted to provide a supportive atmosphere for each other, the Society for Developmental Biology (SDB) had the resources and the resourceful women to make collective consciousness—and collective action—possible. For example, Winifred Doane, who served as one of the officers of the Women's Caucus of the SDB, writes that the SDB had "the acceptance of women members on a par with men, e.g. women were included among the officers and chairs of committees as well as given equal visibility in terms of platform presentations at the symposia. This went as far back as the early 1960s, even before the women's movement got underway. . . . I felt that other women of the SDB were very supportive at times when I really needed some moral support." [56] Biologist-turned-science-studies-scholar Donna Haraway remembers this group in the early 1970s as being proactively feminist and later becoming more career oriented. The members of the Women's Caucus included Mary Clutter, who is now assistant director of biological sciences at the National Science Foundation and who has been very active in the Association for Women in Science (AWIS). She became influential in the development and maintenance of policies aimed at attracting and retaining women in biological sciences. Still another member of this group was Susan Goldhor, who wrote a pamphlet entitled "How to Get a Job," which was distributed at the SDB meetings. It is still useful as an eye-opener for naive graduate students. Dorothy Skinner, Elizabeth Hay, Sheila Counce, Virginia Walbot, and Marie Di Berardino were also prominent biologists who were members of the caucus. These names will be familiar to developmental biologists. They constitute another formidable cross section of the field.

Besides getting more women into developmental biology, or into more prominent positions in this field, feminism was also an important resource for mounting a successful project to transform the nature of what counted as knowledge. The first of these projects was an effort to degender the vocabulary of developmental biology. In this critique, the scientific data themselves have not been questioned so much as the types of questions thought important and the interpretations drawn from experiments and observations. Several individuals and groups have scrutinized this area and have written excellent critiques of its language, its narratives, and its interactions with society.[57] Many of the individuals who have written feminist critiques of developmental biology are themselves trained as biologists. Thus, developmental biology has seen a remarkable reform-from-within. In almost all of these instances, feminist critiques were used in an attempt to make the science "better." Feminists' critiques were used to control for social assumptions and were seen as a tool to bring interpretations back in line with the scientific data. Just as a scientist would control for temperature, pressure, and solvent effects, so the scientist should also control for social biases and cultural assumptions. The Biology and Gender Study Group has called this "controlling for social biases"; Sandra Harding calls it "strong objectivity."[58]

In the last twenty years these critiques have been particularly visible. For example, the Biology and Gender Study Group looked specifically at stories of fertilization and how the sperm and the nucleus are given masculine attributes while the egg and the cytoplasm are made to stand for women. Emily Martin looked at the language being used to describe menstruation, oogenesis, and spermatogenesis, and she came to similar conclusions about how cells became surrogates for men and women. Ruth Hubbard, the Biology and Gender Study Group, and Evelyn Keller have criticized the language being used to represent the genetic mastery over the cytoplasm.

But this program for purging sexist language from developmental biology may be traced back much earlier, to the founding years of the Women's Caucus of the SDB. In 1976, this group published a remarkable pamphlet called *Sexisms Satirized*. As its preface states, "It was made possible through the generous contributions of material from SDB members of both genders. . . . Vexed by recent statements in the biological literature which had sexist overtones, the Women's Caucus decided that satire would be the most effective approach to counteract such remarks. Hopefully the authors quoted here will be persuaded to reassess their objectivity in future publications and the awareness of scientists in general will be somewhat heightened."[59] This pamphlet is noteworthy for many reasons. First, it is one of the very earliest feminist critiques of biology, written before the better-known early analyses of Haraway, Bleier, Hubbard and colleagues, and Gilbert.[60] It is

"In all systems that we have considered, maleness means mastery; the Y-chromosome over the X, the medulla over the cortex, androgen over oestrogen. So physiologically speaking, there is no justification for believing in the equality of the sexes; *Vive la différence!*"

Figure 4.2 Cartoon from *Sexisms Satirized,* published by the Women's Caucus of the Society for Developmental Biology, 1976. The quotation being satirized is from R. V. Short in *Reproduction in Mammals,* book 2, page 70, C. R. Austin and R. V. Short, eds. Reproduced courtesy of the Society for Developmental Biology.

even antecedent to Pauline Bart's 1977 chapter in *Biology as a Social Weapon.*[61] Second, this critique of biology uses irony to make its point. It does not give a philosophical justification or an exposition on the roles of gender in science. Rather, it just quotes *verbatim* the offending text and uses a cartoon to illustrate the point. Figure 4.2 shows one example of the material in this book. This example also demonstrates the type of sexism present in some areas of developmental biology. Third, this pamphlet is a collaboration among women and men; the inclusion of men in the formulation of feminist critiques has been characteristic of developmental biology.[62] Fourth, this was an internal critique, written by scientists for scientists. The critique was couched in friendly terms and was done in the name of better science. This also became a characteristic of the feminist critiques of developmental biology.

The second feminist project in developmental biology has been to perform critiques on various research programs. Again, like the language critique, this project is being done largely within developmental biology by developmental biologists, and it seeks to bring the rhetoric of scientific programs back in line with what the data indicate. One of the most important of these programs has been the critique by Eva Eicher and Linda Washburn of sex determination stories.[63] They pointed out that the standard story was that the default state of sex determination was femaleness, and that maleness was femaleness with something extra. This idea went right back to Aristotle's notion of sex determination that saw females as incomplete men. (And this notion was parodied in the *Sexisms Satirized* brochure.) Eicher and Washburn showed that this story was believable only if one confused primary and secondary sex determination. If you castrate a mammalian embryo, its phenotype becomes female. But that is secondary sex determination and has nothing to do with whether the bipotential gonad rudiment becomes a testis or an ovary. Primary sex determination is actually a bifurcating path, and both testis and ovary formation are active, gene-directed events. However, because of the earlier confusion, "sex determination" was almost entirely synonymous with "male determination," and the scientific research program was to identify testis-forming genes. Ovary-forming genes were not looked for until the 1990s, and two have recently been discovered. Feminist critiques of specific areas in developmental biology have also criticized certain research programs in hormones and brain development (Ruth Bleier, Anne Fausto-Sterling) and molecular biology (C. H. Waddington, Stuart Newman, Brian Goodwin, Ruth Hubbard, Evelyn Fox Keller, Bonnie Spanier). Numerous men are involved in these critiques, and each of these critiques has been advanced in the name of making the science more rigorous.

Both these programs—to change the vocabulary of the discipline and to criticize research programs that have bent science to social norms—have had large, although not complete, success, and these critiques have made their way into the teaching literature of the field. For example, one of us (SFG) writes a mainstream textbook which has been widely used in the field for the past decade. This text refers to and makes use of the above-mentioned critiques of biology as well as the analyses of science studies scholars such as Londa Schiebinger, Susan Bell, Donna Haraway, and Cor Van de Weele.[64] In the pamphlet *From Egg to Adult,* published by the Howard Hughes Medical Institute, the interactions between sperm and egg are described as a dialogue wherein the egg is seen as an active participant in the fertilization process. Similarly, the article on sex determination in this pamphlet states explicitly, "Becoming female is not a default pathway."[65]

CONCLUSIONS

The question we began with was, Have women made developmental biology a "feminist science"—or has feminism changed the means by which we do developmental biology in other ways? Our preliminary answer is a qualified "yes" to both aspects. Certainly, the main agenda of the Women's Caucus of the SDB has been met. Women scientists are no longer confronted with the expectation that the highest rank to which they can reasonably aspire is that of a senior research associate. This success, of course, is not peculiar to developmental biology. However, it certainly can be said to be due to feminism in professional terms, because particular women actively fought for policy changes in funding and representation.[66] Thanks to policies at the National Science Foundation (developed by scientists such as Mary Clutter of the Women's Caucus of the SDB), women became more prominent at meetings, and women were able to present their research more visibly. In developmental biology, there was no problem in finding women to chair sessions and give plenary sessions. Some of the most well known investigators in the field have been women. In several instances, the status of women scientists changed from one of "soft money" to one of tenure track, following the foundation's recognition of their scientific contributions.[67]

In the SDB, "once it became clear that the Society truly did support its women members, the need for the caucus evaporated. . . . Betty Hay became president of the Society and continued the tradition of supporting its women members."[68] But the equality that had been envisioned did not materialize, and the SDB formulated panels to deal with "Women's Issues." These mainstream panels have often been aimed at equalizing the practical education that men might be given by mentors in areas of negotiation, campus politics, and grant writing. In one session (held before a packed auditorium at the University of Wisconsin, Madison), graduate students and postdoctoral fellows (of both genders) were taught how to negotiate and what to expect. It was brought up that this was something that men often were told, but until recently women were just made to feel grateful for having been accepted by the university at all. One woman postdoctoral fellow urged women to be sure to negotiate for a parking space close to the laboratory, because the new recruit could expect to be there at weird hours. At another recent SDB meeting, University of Michigan professor Kathryn Tosney was given a major evening lecture session to explain the "rules behind the rules" of tenure. What is important is that these sessions were not considered "extra," nor were they expected to be attended only by women.

In a related project, women have been protesting against the current state of tenure evaluation, wherein a woman is expected to produce her best science precisely in those years when she might be raising a young family. Princeton developmental geneticist Shirley Tilghman has been the most articulate spokesperson for that cause, and her essays in the *New York Times* sparked great debate and perhaps even some changes.[69] She relates these changes to the numbers of women entering the field, noting: "There is only one solution and that is the recruitment of more women into science. Numbers really matter. When women reach a critical mass, the cultural barriers naturally begin to slip away."[70]

We believe that feminism has indeed made a difference in developmental biology in several ways. First, large numbers of women have not only entered the field but have become its exemplars both scientifically and professionally. Feminism is challenging the politics of science more broadly and the ways in which hiring and promotion are done, and these changes have been incorporated into developmental biology in many prominent ways. Second, feminism has at least partially succeeded in changing the knowledge produced. The vocabulary of the field has been transformed, resulting in a less sexist, less culturally biased, and more scientifically congruent view of the world. Further, this shift has challenged and in some cases changed the ways the field's practitioners have viewed sex determination, fertilization, and brain development. If feminism succeeds in its internal critique of the discipline, this will be an important success, because developmental biology, like primatology, is in the business of telling us who we are and how we came to be. If it succeeds in changing the politics of science, this will also be important, for as Tilghman has written: "The reason we care so much about this subject is that science is an extraordinary field. I know of few other professions where the excitement that brought you to the field is sustained over so many years. It would be a tragedy to exclude women from all this fun."[71]

We also envision ways in which this transformed developmental biology can inform how we understand its history. Methodologically, developmental biology recognizes that what works for one organism may not work for a closely related organism: no one scheme explains all the data. Similarly, what makes one woman a scientist is not necessarily what would make another woman a scientist; what is an active agent in one set of circumstances may be poisonous in another time or place. Developmental biology also teaches us that in the determination of mammalian cell fate, context is critical. Whether a cell becomes a skin cell or a nerve cell, cartilage or muscle, is determined by the other cells it meets. A cell is not intrinsically programmed. Who we meet—our friends, our parents, our teachers—are critical. Finally, developmental biology recognizes what it calls a "community effect": numbers matter.[72] Groups can respond to stimuli differently from isolated individuals. This is also important for historians to consider.

We believe, however, that more individual case studies (along some of the lines we suggested) will be broadly instructive for generating historical comparisons that help explain the cultural symbolic meaning and power of gender in this field. Waelsch and Nüsslein-Volhard, for example, are two very different people: the only real constant in their situations was being perceived as women. Such a perception endows one with a certain recognition of one's body, one's society—and of certain privileges and certain constraints, though these differ from place to place, time to time, household to household, laboratory to laboratory. Multiple situatedness also works in different ways at different times. Where political and social upheavals permitted the German woman Nüsslein-Volhard to be trained in particular areas of biology and to act where she felt she could make the most meaningful contributions to developmental biology, different political and social upheavals constrained the German Jewish woman Waelsch to be at the intersection of mammalian development and genetics, a place where she could work but where few other people were working. Nevertheless, as a result of the more recent feminist critiques, the contemporary culture of developmental biology brings its own perceptions of women and gender to bear on her intervention. Waelsch's article "In Praise of Complexity" thus becomes evidence that having more women in the field makes for different science. In this way, gender clearly impacts community understandings and community behavior—for both scientists and historians of science. From a historian's perspective, then, understanding this effect is most important for understanding how the stories we write about developmental biology's past simultaneously reflect and shape our understandings of the roles women and feminism should play in its future.[73]

NOTES

We wish to thank Dr. Winifred W. Doane for her letters, her insights, and her support; Dr. Ida Chow, business manager of the Society for Developmental Biology, and all those researchers who answered our questions concerning the role of the Women's Caucus of the SDB; Dr. Thomas Vogt, whose scientific vision and historical determination helped make possible a valuable new oral history of Salome Waelsch and Anne McLaren; and, finally, the editors of this volume, for a stimulating symposium at Princeton University and the constructive scholarly exchanges that developed from it.

1. Evelyn Fox Keller, "Developmental Biology as a Feminist Cause?" *Osiris* 12 (1997): 18–19.

2. Ibid., 18.

3. Ibid., 23.

4. Ibid.

5. Ibid., 24.

6. The distinction between developmental biology and embryology is a very loose one. The term "developmental biology" can be said to have originated twice. The first time was in the 1950s, when John Berrill and Paul Weiss introduced it for the title of Weiss's journal, *Developmental Biology*. The term was meant to connote the fact that development includes more than embryology. It also includes the regeneration, the colonial animal development, and other developmental processes that occur in the adult (such as blood formation). The second time was in the mid-1960s and occurred for the opposite reasons. This time the term was meant to integrate embryology with cellular and molecular biology. The term acquired popularity through the serial *Current Topics in Developmental Biology*, which was expressly devoted to a cell and molecular approach to development. The Society for Developmental Biology was called the Society for the Study of Development and Growth until 1965. See Scott F. Gilbert, *A Conceptual History of Modern Embryology* (New York: Plenum Press, 1991), ix.

7. Diane Paul, *Controlling Human Heredity: 1865 to the Present* (Atlantic Highlands, NJ: Humanities Press, 1995), 55.

8. Margaret W. Rossiter, *Women Scientists in America: Struggles and Strategies to 1940* (Baltimore, MD: Johns Hopkins University Press, 1982).

9. See Jane Maienschein, *One Hundred Years Exploring Life, 1888–1988* (Boston: Jones and Bartlett, 1989).

10. Garland Allen, *Life Science in the Twentieth Century* (New York: John Wiley and Sons, 1975), chap. 1.

11. See Maienschein, *One Hundred Years,* 157.

12. Robert E. Kohler, *Lords of the Fly: Drosophila and the Experimental Life* (Chicago: University of Chicago Press, 1994), 96.

13. See Karen A. Rader, "The Mouse People: Murine Genetics Work at the Bussey Institution, 1909–1936," *Journal of the History of Biology* 31 (1998): 327–54.

14. Amy Sue Bix, "Experiences and Voices of Eugenics Field-Workers: 'Women's Work' in Biology," *Social Studies of Science* 27 (1997): 625–68.

15. See Scott F. Gilbert and Marion Faber, "Looking at Embryos: The Visual and Conceptual Aesthetics of Emerging Form," in *The Elusive Synthesis: Aesthetics and Science,* ed. Alfred I. Tauber (Dordecht: Kluwer Academic Publishers, 1996), 125–51; and Scott F. Gilbert, "Bearing Crosses: The Historiography of Genetics and Embryology," *American Journal of Medical Genetics* 76 (1998): 168–82.

16. C. H. Waddington, *New Patterns in Genetics and Development* (New York: Columbia University Press, 1962).

17. Rossiter, *Women Scientists in America.*

18. Transcript of Oral History Interview with Salome Waelsch and Anne McLaren, by Karen Rader and Thomas Vogt, Princeton, NJ, June 1997 (hereafter OHI/97), 8.

19. Salome Glueckson Waelsch, "The Causal Analysis of Development in the Past Half Century: A Personal History," *Journal of Development* supplement (1992): 1.

20. Salome Glueckson Waelsch, "The Development of Creativity," *Creativity Research Journal* 7 (1994): 249.

21. OHI/97, 41.
22. Ibid., 54.
23. Ibid., 4
24. Waelsch, "Causal Analysis of Development," 1.
25. See Waelsch "Causal Analysis of Development" and OHI/97.
26. OHI/97, 50.
27. Waelsch, "Causal Analysis of Development," 1.
28. Ibid., 2; OHI/97, 45; Salome Gluecksohn Waelsch, "Viktor Hamburger and Dynamic Concepts of Developmental Genetics," in *Studies in Developmental Neurobiology: Essays in Honor of Viktor Hamburger,* ed. W. Maxwell Cowan (New York: Oxford University Press, 1981), 44–52.
29. Waelsch, "Development of Creativity," 250.
30. Waelsch, "Causal Analysis of Development," 2.
31. Edward Yoxen, "Form and Strategy in Biology: Reflections on the Career of C. H. Waddington," in *A History of Embryology,* ed. T. J. Horder, J. Witkowski, and C. C. Wylie (Cambridge: Cambridge University Press, 1986), 309–29, quotation on 311.
32. C. H. Waddington, *Organisers and Genes* (Cambridge: Cambridge University Press, 1940), 1. For more on this issue, see Scott F. Gilbert, "Induction and the Origins of Developmental Genetics," in Gilbert, *A Conceptual History of Modern Embryology,* 181–206; and Scott F. Gilbert, "Epigenetic Landscaping: Waddington's Use of Cell Fate Bifurcation Diagrams," *Biology and Philosophy* 6 (1991): 135–54.
33. Waelsch, "Development of Creativity," 250.
34. OHI/97, 60.
35. Ibid., 63.
36. Ibid., 70, 73.
37. Ibid., 199 (see also McLaren's comments on the same page).
38. Ibid., 72.
39. Ibid., 76.
40. Ibid.
41. See Gilbert, "Induction and the Origins of Developmental Genetics."
42. Waelsch, "Causal Analysis of Development"; see also S. Waelsch, "In Praise of Complexity," *Genetics* 122 (Aug. 1989): 721–25.
43. Waelsch, "In Praise of Complexity," 721.
44. See also Gilbert, "Induction and the Origins of Developmental Genetics."
45. OHI/97, 166.
46. Interview with Salome Waelsch in *The Outer Circle: Women in the Scientific Community,* ed. Harriet Zuckerman, Jonathan R. Cole, and John T. Bruer (New York: W. W. Norton, 1991), 71–93.
47. C. H. Waddington, "The Practical Consequences of Metaphysical Beliefs on a Biologist's Work: An Autobiographical Note," in *The Evolution of an Evolutionist* (Ithaca, NY: Cornell University Press, 1975), 3.
48. Waddington, *Organisers and Genes,* 3–4; see also Gilbert, "Induction and the Origins of Developmental Genetics"; and Gilbert, "Epigenetic Landscaping."
49. Yoxen, "Form and Strategy," 323–24. Early on, Waddington himself seems to have been particularly sensitive to the marginalized position of women in biology. In a letter asking Theodosius Dobzhansky to temper some of his criticisms of a particular woman geneticist, Waddington wrote that neither Waelsch nor Barbara McClintock had an official position in line with their scientific abilities or efforts: "I think that women biologists in America have in any case a very difficult job to get themselves accepted—look at Barbara McClintock, as the most extreme case of an absolutely first-rate person who has been forced into the position of an eccentric recluse; and the positions of Salome Waelsch, Jane Oppenheimer, Dorothea Rudnick is only slightly better" (letters from Waddington to

Dobzhansky, Mar. 11 and 20, 1964 [with Dobzhansky's letter to Waddington appended, Mar. 15, 1964], Dobzhansky Papers, American Philosophical Society).

50. See Biology and Gender Study Group, "The Importance of Feminist Critique for Contemporary Cell Biology," *Hypatia* 3 (1988): 61–76.

51. Gilbert and Faber, "Looking at Embryos."

52. James W. McAllister, *Beauty and Revolution in Science* (Ithaca, NY: Cornell University Press, 1996).

53. Ronald Rainger, Keith Benson, and Jane Maienschein, eds., introduction to *The American Development of Biology* (Philadelphia: University of Pennsylvania Press, 1988), 3–11.

54. Zuckerman, Cole, and Bruer, *The Outer Circle*, 82.

55. Keller, "Developmental Biology as a Feminist Cause?" 23.

56. Winifred W. Doane, letter to SFG, Jan. 25, 1999.

57. Donna J. Haraway, *Crystals, Fabrics, and Fields: Metaphors of Organicism in Twentieth-Century Developmental Biology* (New Haven, CT: Yale University Press, 1976); Ruth Hubbard, Mary Sue Henifin, and Barbara Fried, eds., *Women Look at Biology Looking at Women* (Cambridge, MA: Schenkman, 1979); Gerald Schatten and Heidi Schatten, "The Energetic Egg," *Sciences* 23 (1983): 28–34; Ruth Bleier, *Science and Gender: A Critique of Biology and Its Theories on Women* (New York: Pergamon Press, 1985); Anne Fausto-Sterling, *Myths of Gender: Biological Theories about Women and Men* (New York: Basic Books, 1985); Eva M. Eicher and Linda Washburn, "Genetic Control of Primary Sex Determination in Mice," *Annual Review of Genetics* 20 (1986): 327–60; Biology and Gender Study Group, "The Importance of Feminist Critique"; Emily Martin, "The Egg and the Sperm: How Science Has Constructed a Romance Based on Stereotypical Male-Female Roles," *Signs: Journal of Women in Culture and Society* 16 (1991): 485–501; Evelyn Fox Keller, *Refiguring Life: Metaphors of Twentieth Century Biology* (New York: Columbia University Press, 1995).

58. Biology and Gender Study Group, "The Importance of Feminist Critique"; Sandra Harding, *Whose Science? Whose Knowledge? Thinking from Women's Lives* (Ithaca, NY: Cornell University Press, 1991).

59. Winifred W. Doane, ed. (cartoons by B. K. Abbott), *Sexisms Satirized* (Pocketbook Profiles/Society for Developmental Biology, 1976).

60. Donna Haraway, "Animal Sociology and a Natural Economy of the Body Politic I and II," *Signs: Journal of Women in Culture and Society* 4 (1978): 21–60; Ruth Bleier, "Social and Political Bias in Science: An Examination of Animal Studies and Their Generalization to Human Behavior and Evolution," in *Genes and Gender II: Pitfalls in Research on Sex and Gender*, ed. Ruth Hubbard and Marian Low (New York: Gordian Press, 1978), 49–69; Hubbard, Henifin, and Fried, *Women Look at Biology;* Scott F. Gilbert, "The Metaphorical Structuring of Social Perceptions," *Soundings* 62 (1979): 166–86.

61. Pauline B. Bart, "Biological Determination and Sexism: Is It All in the Ovaries?" in *Biology as a Social Weapon,* ed. Ann Arbor Science for the People Editorial Collective (Minneapolis: Burgess, 1977), 69–83.

62. SFG, for example, received his M.A. in the history of science under the aegis of Donna Haraway, and his Ph.D. in biology in the laboratory of Barbara Migeon, two very different types of feminists.

63. Eicher and Washburn, "Genetic Control of Primary Sex Determination in Mice."

64. S. F. Gilbert, *Developmental Biology,* 6th ed. (Sunderland, MA: Sinauer Associates, 2000).

65. Maya Pines, ed., *From Egg to Adult: What Worms, Flies, and Other Creatures Can Teach Us about the Switches That Control Human Development* (Bethesda, MD: Howard Hughes Medical Institute, 1992).

66. Political mentors were also important in this process. Heinrich Waelsch,

Salome's second husband, not only was "extremely cooperative and helpful" regarding her desire to have children while continuing to pursue her T-locus work (OHI/97, 87) but also was her political mentor. By her own recollection, it was "Heini" who crystallized her resolve to approach Dunn about professional advancement at Columbia, given the vast amount of work she had accomplished (Zuckerman, Cole, and Bruer, *The Outer Circle,* 82–83). He also helped her use her newfound fame to get Columbia president Dwight D. Eisenhower to obtain a larger university apartment for them (OHI/97, 108–9).

67. M. E. Clutter, letter to SFG, Mar. 8, 1999.

68. Winifred W. Doane, letter to SFG, 1999.

69. Shirley M. Tilghman, "Science vs. the Female Scientist," Op-ed, *New York Times,* Jan. 25, 1993, sec. A, p. 17, col. 1; Shirley M. Tilghman, "Science vs. Women—A Radical Solution," Op-ed, *New York Times,* Jan. 26, 1993, sec. A, p. 23, col. 2.

70. Tilghman, "Science vs. Women."

71. Ibid. Tilghman, an outspoken feminist, was elected president of Princeton University in 2001.

72. J. B. Gurdon, P. Lemaire, and K. Kato, "Community Effects and Related Phenomena in Development," *Cell* 75 (1993): 831–34.

73. See also Margaret W. Rossiter, "The Matilda Effect in Science," *Social Studies of Science* 23 (1993): 425–41.

FIVE

Making a Difference: Feminist Movement and Feminist Critiques of Science

EVELYN FOX KELLER

Twenty-five years ago, the terms "gender" and "science" had not yet been formally conjoined (at least not in respectable academic circles); nor had the implications of such a conjunction yet been subjected to any kind of serious analytic or historical scrutiny. Over the intervening quarter of a century, however, "gender and science" has become the name of a respected academic subdiscipline: a well-established category of science studies with its own conferences, its own courses, a voluminous body of published literature, and now, as well, its own retrospective volume designed to examine the impact this academic subspecialty has had. Before beginning my reflections on the particular question of the present volume—What difference has feminism made?—I want to make a few observations about the scope of issues subsumed under the label "gender and science."

Elsewhere I have suggested that this label was originally invoked to subsume (and contain) emerging scholarship along three different lines of inquiry (on women in science, on scientific constructions of gender, and on the influence of gender in historical constructions of science).[1] Today, however, even as expansive (and perhaps unwieldy) a definition as this seems inadequate. As the contributions to this volume so clearly illustrate, the referents of the composite term "gender and science" have proliferated profusely, making for an extraordinary variety of avenues of scholarly exploration. Clearly, we have gotten hold of a gargantuan entity—not only huge but amorphous and ever changing, going off in so many different directions, and raising so many different kinds of issues that it may well be that the only thing that holds this entity together is Helen Longino's bottom-line axiom, that is, its active resistance to the disappearance of gender (as well as, of course, of women).[2] But I do not want to be misunderstood: I regard the growth and perfusion of the subject as an unambiguously positive achievement. Indeed, in some ways, it represents the realization of a goal that has been central to feminist studies from their earliest days, namely, that of integrating the insights and methods of feminist scholarship with other, more conventionally familiar academic concerns. Still, it should be acknowledged that such proliferation of meanings does create certain difficulties in dealing with the question that has been put to us here. My strategy for dealing with these difficulties is to begin by attempting to parse that very question. I want to ask, What do we mean by "making a difference"? What do we mean by "feminism"? And just what kind of relation obtains between feminism and the new scholarship on gender and science?

MAKING A DIFFERENCE OR, RATHER, MAKING A DIFFERENCE TO WHAT?

The question, What difference has feminism made? is manifestly incomplete: it points clearly enough to a direct object (the specified *what*, i.e., "difference"), but it omits or suppresses the requisite indirect object (the unspecified *to what*). Necessarily, when we speak of making a difference, we have in mind one or more particular arenas of interest in which a difference might be made. To take a simple example, it is apparent to everyone that feminism has made a substantial difference to many of our careers, but presumably, that difference is not the subject of interest here. What kinds of difference, then, *are* of interest? What do we have in mind when we ask this question?

It is not customary for scholars to inquire about the purpose of their work, to ask what the production of scholarship is *for*. Apart from the immediate and manifestly interested goal of the furtherance of careers, the aim is assumed to be both disinterested and self-evident: for example, truth; an increase in knowledge; or, simply, better understanding. The very fact that such a question is now being put to feminist scholarship can be seen as a consequence of the rootedness of this scholarship in another kind of endeavor altogether, one aimed, not at the growth of knowledge per se, but at political transformation. Indeed, in its early days, the very engagement in feminist scholarship was seen as requiring a rationale, an answer to the essential tension then so widely felt between *feminism* on the one hand and *scholarship* on the other. That rationale lay in the claim for knowledge as itself constituting a tool for social transformation, for knowledge as "politics by other means." But perhaps it was inevitable that as the new scholarship on women and gender gained in academic respectability, the arena of envisaged transformation narrowed steadily: if knowledge is politics by other means, then that politics came progressively to be understood as intellectual politics (as a politics of knowledge), and the social world most immediately envisaged as the object of transformation became that of the academy.

Thus it is that one of the most obvious arenas of interest—and a manifest focus of much if not most of the recent scholarship on gender and science—came to be, after all, the familiar arena of scholarship, with transformation pursued through the enhancement of knowledge and understanding. But even so construed, we are again left with a missing object, for we need to ask, Knowledge and understanding of what? And it is in response to this last question that the chapters of this volume most conspicuously divide, roughly along the following lines:

1. For some authors, an arena of obvious interest is our understanding of women's experiences. These efforts might be subsumed under the larger project of "writing women back into history" that was originally launched in the 1970s with the clear expectation that knowledge of the past would empower women in the present and, as such, help to transform women's lives. Today, of course, expectations have shifted. But however much the indirect aim of large-scale social transformation has receded from the agenda, the more direct aim of understanding women's experiences has remained constant in much contemporary scholarship. Accordingly, and perhaps inevitably, recent efforts invite a similar kind of additional fracturing. For as we have learned, it makes dubious sense to ask about women's experiences without first asking, and quickly, Which women?

Bifurcations inscribed by questions of race, class, ethnicity, and sexuality are, of course, the most familiar. But for work in gender and science (or, more generally, in gender and science, technology, and medicine [GSTM]), I would argue that a rather different set of distinctions—to be sure, crosscutting and interweaving these more familiar categories—takes priority. Science (as well as technology and medicine) interfaces with the lives of women in many different ways and from a variety of different subject positions that are only secondarily marked by matters of racial, economic, or sexual identity. Very crudely, I would distinguish these different positions as follows:

- women as subjects (or objects) of scientific, technological, and medical research;

- women as members (or would-be members) of professional scientific, technological, or medical communities;
- women as users or consumers of scientific, technological, or medical developments.

In this framework, the question, Which women? directs us to a primary grouping (and division) of women according to the risks and benefits proffered by their particular kind of engagement with science, technology, or medicine. And it is obvious that the kinds of risks and benefits encountered in these different subject positions diverge radically. Indeed, it is just because of such radical divergences that I would argue both for the usefulness of such a distinction in categorizing research in this arena and for the need to make such a distinction at the start of any inquiry into an interaction between science (or technology or medicine) and the experiences of women.

2. Another substantial body of scholarship in GSTM points elsewhere and is aimed at a better understanding, not of women's experiences per se, but of the role of dominant ideologies of gender in scientific, technological, and medical history. The direct beneficiaries of such analyses have been science studies, that is, the history, sociology, and even philosophy of science, technology, and medicine. To paraphrase Alison Wylie, these analyses have served to throw into relief a range of largely unrecognized assumptions and, in some cases, associated patterns of practice that have profoundly shaped inquiry.[3] Accordingly, they have contributed to new understandings of the scientific revolution, of the history of systematics, of nineteenth-century craniometry, of twentieth-century primatology, developmental biology, archeology, and so forth.[4] The list is long.

3. A third body of scholarship has focused on changing cultural maps—for example, of sex and gender, of women, of the body—and has helped give rise to such new arenas of research (e.g., the history of the body, histories and anthropologies of sex and gender) as comprise the rapidly growing field of cultural studies of science.[5] In all these arenas, the impact of feminist scholarship has been both undeniable and conspicuous.

4. Yet a fourth body of scholarship shifts the focus once again, turning our question into a reflexive one: this scholarship obliges us to ask not only "which women" but also "which feminism and which feminists" are we speaking of? Mindful of this very ambiguity, bell hooks once offered a remarkably felicitous locution: she urged us to abandon talk of "feminism" or "feminists" altogether and to speak instead of "feminist movement."[6] Hooks's term, "feminist move-

ment," extends a twofold invitation: first, to recognize and, second, to embrace the multiple and changing character of feminism; as such, it restores to that word the dynamic connotation that is, after all, only appropriate to a historical social movement. It also suggests another phrasing of our original question, namely, What difference has feminist movement made to studies in gender and science?

Such a rephrasing prompts some further observations. Feminist movement has made an obvious and dramatic difference to the subject of gender and science by compelling feminist scholars to reflect on their own tacit assumptions, that is, on our own kinds of myopia. And such self-reflection (sometimes self-flagellation) has led to the identification of a number of lacunae in the first round of feminist analyses (of which there were many). Most notoriously, it has promoted sensitivity to and sophistication about issues of race and its relation to gender and also to sexuality. Here feminist movement has served as a corrective, helping to enlarge our understanding, not only of women's experience as subjects and objects of science, technology, and medicine, but also as professionals and consumers. It might also be noted that feminist movement has changed the constituency of the gender and science community. The first generation of scholars in this field tended to be women, and white women at that. Also, many of them had originally been trained as scientists, and were indeed often still active and enthusiastic participants in scientific research. Today, the subject has at least partially escaped the confines of these narrow categories (even if more visibly on some boundaries than on others): although this community is still largely white and female, it seems to be growing more diverse in its gender and ethnic makeup, and it is certainly growing more diverse in its disciplinary makeup. Indeed, it might be said that one of the casualties of feminist movement has been the diminished participation of women scientists and the consequent erasure of the passion (however ambivalent) that they had once brought to this subject.

Finally, there is quite another kind of difference we might also ask about, an altogether different way of reading the original question, and that is as a question about the difference feminism (or feminist scholarship) has made to the practice of science, technology, or medicine (and hence, perhaps, to our understanding of natural phenomena). Such a reading of the question becomes particularly relevant if we recall that just as there was a political subtext to the early feminist scholarship on women, so too there was a political subtext to much of the early scholarship on gender and science—aimed as much at "liberation" of science as at the liberation of women. Indeed, a central aim of my own early work was, as I wrote in 1985, to "clarify part of the substructure of science in order to preserve the things that science has taught us, in order to be more objective."[7] But by the tenth-year anniversary reprinting of my book, I had become notably more circumspect in my ambitions. It was evident to me that feminist scholarship in gender and science had had a substantial impact on the history, sociology, and anthropology of science—and a rather smaller (but still noticeable) impact on the philosophy of science, but as to the practice of science itself, I was much less sure. Here, however, I want to modify my claim and to argue for the possibility of identifying a clear impact on the practice of science that has been made over the past two decades by feminism, even if not by the actual literature of gender and science or even, for that matter, by feminist scholarship more generally. However, to make this argument, I need first to draw attention to a certain chronic slippage that is in fact invited by the very posing (i.e., syntax) of our original question, namely, the slippage between "feminists" and "feminism."

WHAT IS IT THAT MAKES A DIFFERENCE?

In the preceding section, I argued for the need to restore the missing indirect object of our question (to what is a difference being made?); here, I want to stress the need to disambiguate its subject (that which makes a difference). The explicit subject is feminism. But making a difference implies agency, and we are not normally accustomed to attributing agency to abstract entities, such as social movements; we suppose that only human actors—ourselves—have agency. Thus it is that so many participants in this volume responded to its organizing question by tacitly reading "feminism" as "feminists," that is, as ourselves, claiming for "us" not only identification with the social movement but causal responsibility for it.

But it seems to me that by far the most important differences we have seen over the last quarter century are not those made by particular individuals or groups but, rather, those that were forged by feminism as a social movement, and here, I unabashedly use the singular form of the term. The particular social movement that we call (modern) feminism acquired its meaning, and its force, not by the platforms or arguments of any of its different variants, but by the glue that bound these variants into an operational unity. Whatever the tensions, the conflicts, the multiplicities that existed within that movement, the important thing is that these tensions, conflicts, and multiplicities could be contained, even submerged, by the greater force of the whole—at least for a while, at least for long enough in any case to mount one of the most dramatic cultural revolutions of recent history.

As Norton Wise reminds us, a hundred years ago, Gustave Le Bon wrote quite disparagingly of phenomena like mass movements and crowds—likening them to such "lower forms of evolution" as "women, savages, and children."[8] What makes them of an inferior order is their mode of operation—crowds and mass movements do not proceed by the conscious will of individuals so much as by the repetition of simplified ideas and images through popular culture, a form of contagion that sweeps through the crowd by imitation. Like so many others, I believe that Le Bon overestimated the power of individual consciousness about as seriously as he underestimated the force of such large-scale social effects as sweeping through the crowd by imitation. I suggest that the impact of feminism on gender and science provides us with a case in point.

For all the difference that feminist scholars have made (and I include myself), however brilliant, forceful, and imaginative our contributions have been, I want to argue that the real agent of change—if you like, the real heroine of the past three decades—has been the social movement itself. Indeed, feminist scholars themselves are now and have from the beginning been a product of that movement, especially in the United States. Of course, it goes both ways, but it is a noteworthy historical fact that, in this country at least, the emergence of feminist scholarship (and, more specifically, the subject of gender and science) was in fact preceded by a political and social movement. To be sure, the feminist movement began with the efforts of a few individuals and groups, but very quickly it took on a life of its own, sweeping into its active center all the cultural machinery of a generation (indeed, that is precisely what made it a social movement). And feminist scholarship was only one of its many by-products. The maelstrom of second-wave feminism gave rise to a men's movement, to a generation of caring fathers, to a profusion of new women detectives (both in novels and on television), to new forms of speech, to new legislation, to new social mores. In a word, it transformed the meaning of gender. And one of the most dramatic by-products of this transformation, especially in the context of gender and science, was the opening of science, engineering, and medicine to women and the dramatic influx of at least white women into these arenas.

And what difference, if any, has this made to the practice of science (or engineering or medicine)? In particular, has it led to any noticeable changes in the content of science? Without question, the influx of women into the profession has changed the profile of American scientists at least. But has it in any observable way changed the actual doing of science? One might argue that changing the face of science does in itself change the doing of science, but not, I believe, in the ways people tend to think of first. What is usually understood by this question is: Have the women themselves changed the doing of science? Have they by their own example brought a new legitimization of traditionally feminine values into the practice of science? Thus posed, my guess would be: probably not. With a few possible exceptions, I do not believe that women scientists have either sought to or succeeded in introducing stereotypic feminine values into the lab—indeed, logic itself seems to me to argue against such a possibility. As the most recent group to be integrated, women scientists are under particular pressure to shed whatever traditional values they may have absorbed qua women—if for no other reason than merely to prove their legitimacy as scientists.

But if we were to rephrase the question and ask, Has their presence helped to restore equity in the symbolic realm in which gender has operated for so many eons? I would answer with an unequivocal yes. Especially, I would argue that the commonplace presence of women in positions of leadership and authority in science has helped erode the meaning of traditional gender labels in the very domain in which they worked, and for everyone working in that domain. Furthermore, I would argue that this erosion has helped to open up new cognitive spaces and has thereby contributed to concrete changes in the very content of at least some scientific disciplines. Let me explain what I mean by giving some examples from the discipline that best illustrates my claim, namely, from biological science.

We might start with the simplest example and, thanks in good part to the work of Emily Martin[9] and Scott Gilbert and his students,[10] the example that is probably best known: that of fertilization. Until fairly recently, the sperm cell has consistently been depicted as "active," "forceful" and "self-propelled," enabling it to "burrow through the egg coat," "penetrate" the egg, to which it "delivers" its genes, and "activate the developmental program." By contrast, the egg cell is passively "transported" or "swept" along the fallopian tube until it is "assaulted," "penetrated," and fertilized by the sperm.[11] The noteworthy point here is not that this is a sexist portrayal (of course it is) but, rather, that the technical details elaborating this picture have been, at least until the last few years, astonishingly consistent with it: the experimental work provided chemical and mechanical accounts for the motility of the sperm, for their adhesion to the cell membrane, and for their ability to effect membrane fusion. The activity of the egg, assumed a priori to be nonexistent, required no mechanism, and no such mechanism was found.

Only recently has this picture shifted, and with that shift, so too has shifted our technical understanding of the molecular dynamics of fertilization. In an early and self-conscious marking of this shift, two researchers in the field, Gerald Schatten and Heidi Schatten, wrote in 1983: "The classic account, current for centuries, has emphasized the sperm's performance and relegated to the egg the supporting role of Sleeping Beauty. . . . The egg is central to this drama, to be sure, but it is as passive a character as the Grimm brothers' princess. Now, it is becoming clear that the egg is not merely a large yolk-filled sphere into which the sperm burrows to endow new life. Rather, recent research suggests the almost heretical view that sperm and egg are mutually active partners." [12] Indeed, the most current research on the subject routinely emphasizes the activity of the egg cell in producing the proteins or molecules necessary for adhesion and penetration. In a recent issue of *Nature,* for example, we can read: "At one time, eggs were regarded like the cargo in the hold of a vessel. . . . We now recognize that each egg actively influences the development of its own follicle— it despatches commands affecting the growth and differentiation of the granulosa cells around it, while receiving information and nutrition from them. . . . This could throw fresh light on unexplained forms of infertility and indicate new strategies for contraception." [13] Even the widely used textbook *Molecular Biology of the Cell* seems to have embraced at least nominal equity on the matter: here, "fertilization" is defined as the process by which egg and sperm "find each other and fuse." [14]

This story provides a powerful lesson. It illustrates the ways in which language can shape the thinking and acting of working scientists—that is, by framing their attention, their perception, and, accordingly, the fields in which they can envision experiments that might be useful to undertake. Of course, not all metaphors are equally productive, but in this particular example, both metaphors were manifestly productive, albeit of different effects: one led to intensive investigation of the molecular mechanisms of sperm activity, and the other fostered research permitting the elucidation of mechanisms by which the egg would have to be said to be "active."

Over the last twenty years, the impact of changing cultural perceptions of gender on the biological sciences has grown steadily, extending well beyond studies of fertilization per se to include a host of research endeavors enriched by a new appreciation of a wide range of phenomena grouped together under the term "maternal effects." "Maternal effects" refer to those long-range influences on the biology of offspring (and even on the evolution of species) resulting from the particular behavior and physiology of the maternal parent. Indeed, the role of the egg in enabling (or initiating) fertilization can be described as a maternal effect, as can (and is) the role of the egg's cytoplasm on the developing zygote. Maternal effects thus arise in a wide range of biological disciplines, from evolutionary biology and ecology to developmental genetics. Because I have written elsewhere about the role of maternal effects in developmental genetics,[15] here I will restrict myself to a discussion of their importance in evolutionary biology and ecology, and I will use two reports in recent issues of *Nature* and *Science* to illustrate.

In a review of William Eberhard's book on sexual selection, T. R. Birkhead writes of changing perceptions of the role of females in evolutionary biology:

> Females have always had a bad deal in evolutionary speculation. When Darwin first proposed the concept of sexual selection, he imagined two processes: competition between males, and female choice. It was obvious that males fought for access to females, but female choice was far from certain, and some people doubted whether females even have the mental ability to make such choices. It has taken more than a century of painstaking research to show that female choice is a subtle but important part of sexual selection.
>
> . . . In 1970, biologists realized that there was more to sexual selection and that, even after copulation and insemination, competition between males could continue, through the process of sperm competition. Inside the passive female's reproductive tract, the sperm from different males grapple for paternity. Most behavioural ecologists were sexist, and it was more than a decade before the complementary idea of cryptic female choice emerged: i.e., that females might also be able to influence the paternity of their offspring. But even this ideal was too early for its time, and was generally thought of as barely credible. . . .
>
> . . . A few years ago, a combination of events changed this perception. . . . By drawing our attention to the many different ways in which females may potentially control paternity, [Eberhard's book] opens up a whole new field of research.[16]

Elisabeth Pennisi has this to say about new developments in ecology:

> [Experimental ecologists have long been] tripped up by a phenomenon that has often been observed but—until recently—rarely seriously considered: a so-called "maternal effect" which occurs when something

about the mother's environment alters how her offspring look, act, and function. . . . Maternal effects . . . are proving to be much more than obstacles that occasionally confound experiments. . . . Maternal effects can enhance an offspring's chances of survival, skew sex ratios, and drive fluctuations in population size. . . .

Researchers had been aware of maternal effects for decades, . . . but for the most part, these early workers viewed these effects as "random noise that tended to obscure the genetic variation that we were interested in," says Mousseau. Thus animal breeders and evolutionary biologists would first grow several generations of the organism they wanted to study in controlled conditions so as to eliminate this "noise."[17]

Clearly, this is an exciting moment in biology—as it is in society. And I am saying that at least in part we have the women's movement to thank. Second-wave feminism has been one of the most powerful social movements of modern times (especially in the United States). Nevertheless, Le Bon's perception of mass movements as "women," as emotional, as antithetical to rigor, argument, and logic, persists. Let me close with a speculation: Perhaps it is this lingering perception of social movements, and feminism especially, as themselves "women" (and, accordingly, as less rational, as lower on whatever intellectual scale) that accounts for the widespread underreading of feminist arguments. I am referring, of course, to the persistent and evidently steadfast proclivity in so much of the academy to reduce these arguments to one or another simplistic essence, to hopeless claims or questions about whether or not women and men think differently—the kinds of claims that have historically always worked to demote the status of women and, hence, the kinds of claims that new generations of women scientists will inevitably rebel against, and rightly so. It is just such a reductive impulse that, I suggest, is what we most vigorously need to resist if we are to protect the gains of the past twenty-five years and guarantee their continuance into the twenty-first century.

NOTES

1. Evelyn Fox Keller, "The Origin, History, and Politics of the Subject Called 'Gender and Science,'" in *Handbook of Science and Technology Studies,* ed. Sheila Jasanoff, Gerald E. Markle, James C. Petersen, and Trevor Pinch (Thousand Oaks, CA: Sage Publishing, 1995), 80–94.

2. Helen Longino, "Cognitive and Non-cognitive Values in Science: Rethinking the Dichotomy," in *Feminism, Science, and Philosophy of Science,* ed. Jack Nelson and Lynn Hankinson Nelson (Boston: Kluwer, 1996), 39–58.

3. This volume, chap. 2.

4. For examples, on the scientific revolution see Evelyn Fox Keller, *Reflections*

on Gender and Science (New Haven, CT: Yale University Press, 1985); on systematics see Londa Schiebinger, "Why Mammals Are Called Mammals," in *Nature's Body: Gender in the Making of Modern Science* (Boston: Beacon Press, 1993), 40–74; on nineteenth-century craniometry see Cynthia Eagle Russett, *Sexual Science: The Victorian Construction of Womanhood* (Cambridge, MA: Harvard University Press, 1989); on twentieth-century primatology see Donna Haraway, *Primate Visions: Gender, Race, and Nature in the World of Modern Science* (New York: Routledge, 1989), and this volume, chap. 3; on developmental biology see Evelyn Fox Keller, "Developmental Biology as a Feminist Cause?" *Osiris* 12 (1997): 16–28, and this volume, chap. 4; and on archeology see this volume, chap. 2.

5. For examples, see Londa Schiebinger, ed., *Feminism and the Body* (Oxford: Oxford University Press, 2000); Thomas Laqueur, *Making Sex: Body and Gender from the Greeks to Freud* (Cambridge, MA: Harvard University Press, 1990); Gail Rubin, "The Traffic in Women: Notes on the 'Political Economy' of Sex," in *Toward an Anthropology of Women,* ed. Rayna Reiter (New York: Monthly Review Press, 1975), 157–220; Gail Rubin, "Sexual Traffic," *Differences* 6 (1994): 62–99; Faye D. Ginsburg and Rayna Rapp, *Conceiving the New World Order* (Berkeley and Los Angeles: University of California Press, 1995).

6. bell hooks, *Feminist Theory: From Margin to Center* (Boston: South End Press, 1984).

7. Keller, *Reflections on Gender and Science,* 178.

8. M. Norton Wise, "Time Gendered and Time Discovered in Victorian Science and Culture," in *From Energy to Information,* ed. Linda Henderson and Bruce Clark (Stanford, CA: Stanford University Press, forthcoming).

9. Emily Martin, "The Egg and the Sperm: How Science Has Constructed a Romance Based on Stereotypical Male-Female Roles," *Signs: Journal of Women in Culture and Society* 16, no. 3 (1991): 485–501.

10. Gender and Biology Study Group, "The Importance of Feminist Critique for Contemporary Cell Biology," in *Feminism and Science,* ed. Nancy Tuana (Bloomington: Indiana University Press, 1987), 172–87.

11. Martin, "The Egg and the Sperm," 489–90.

12. Gerald Schatten and Heidi Schatten, "The Energetic Egg," *Sciences* 23, no. 5 (Oct. 1983): 29.

13. Roger Gosden, "The Vocabulary of the Egg," *Nature* 383 (Oct. 10, 1996): 485–86.

14. Bruce Alberts et al., *Molecular Biology of the Cell* (New York: Garland Press, 1990), 868.

15. See, for example, Evelyn Fox Keller, *Refiguring Life: Metaphors of Twentieth Century Biology* (New York: Columbia University Press, 1995).

16. T. R. Birkhead, "In It for the Eggs," review of *Female Control: Sexual Selection by Cryptic Female Choice,* by William G. Eberhard, *Nature* 382 (Aug. 29, 1996): 772.

17. Elisabeth Pennisi, "A New Look at Maternal Guidance," *Science* 273 (Sept. 6, 1996): 1334–36.

Technology

SIX

Feminism and the Rethinking of the History of Technology

CARROLL PURSELL

As the history of technology professionalized in the 1950s and 1960s, it looked much like the people and things that it studied. Based largely on a foundation of engineering thought and practice, and often undertaken by people with engineering training, the field not only was staffed largely with male scholars but focused on those narrow areas of technology that (white) men had pretty much kept to themselves: invention and engineering design. Biographies of Robert Fulton, Samuel Morse, Thomas Edison, and Cyrus McCormick joined stories of the invention of the steam engine and the designs of the early canals. In 1976 Ruth Schwartz Cowan, bringing the history of technology into contact with the field of women's history, asked, "Was the female experience of technological change significantly different from the male experience?" Six years later in the pages of *Signs* Judith McGaw could already report a "rich, diverse, scattered, uneven, and amorphous body of literature" that dealt in some way with women's experience.[1] Much, though of course not all, of this early literature was, perhaps inevitably, recuperative. For example, work in these years showed that despite all the odds, a few women had indeed received engineering training over the years and that hundreds of women had been granted patents for new inventions.

By 1989 McGaw introduced an even more important advance in the literature.[2] Feminist scholars in other fields had pointed out the overwhelming importance of gender analysis. Using the powerful insights provided by feminist theory, historians of technology could now see the presumptive masculinity of technology not just as a block to the documentation of female participation but as something socially constructed which itself had meanings that could be read. Masculinist attempts to appropriate and defend technical skills, for example, invoke women, workers, and "colored" people as categories of others who do not share those skills: the cultural marginality of these people has been appropriated to stigmatize the unskilled and therefore enhance the power of the skilled. It is now possible to notice the ubiquitous use of the term "manly" when technology is discussed.

Not surprisingly, the growing influence of feminist analysis on the history of technology has made it imperative that we take another look at the very categories we have inherited from the past, which by and large continue to shape our inquiries and our understandings of what we discover. Race and class have also been undertheorized in the field, but taking theory seriously in all three areas has begun to reveal much that we wish to know.

Since, historically, a major effort was made to bar women (and other groups) from the manly pursuit of invention and engineering, they must be looked for elsewhere. When we do that, we discover that the *design* of technology is only one aspect of the matter, and that when we take such activities as acquisition, maintenance, repair, use, and redesign seriously, women, children, workers, and "people of color" reappear in all their diversity and importance. The same is true, of course, when we consider cultural appropriation and interpretation.

In this chapter my intention is to expand upon and give some examples of each of these stages in the feminist rethinking of the history of technology. These remarks will be largely limited to the literature that has originated in the United States, since that is what I know best. I will also address the thorny question of what, if any, effect feminism has had on the *practice* of technology. The answer to that question, I will suggest, flows directly out of the changes taking place in the way we think about the history of women and technology. As usual, it is the questions which prove to be of critical importance, not just the answers.

The history of technology as a separate field, or at least the Society for the History of Technology (SHOT), was established in 1958, less than ten years after the Society of Women Engineers (SWE) was founded in the United States, and it was to be another decade before the nation's engineering schools made a concerted effort to recruit women for their undergraduate programs. Engineering was famously gendered masculine, with perhaps the lowest percentage of female participation of any of the major professions, and it was precisely out of these schools that the serious academic interest in the history of technology arose. Many of the first members of SHOT were trained initially as engineers, many of them taught at engineering schools, and in the early years, it was considered important that every other president of the society be a practicing engineer. These circumstances heavily influenced not only the sex ratio of practitioners but also the topics and questions that were considered interesting and legitimate. As in the history of science and the history of medicine, practitioners were vetted both for a knowledge of the approved subject matter but also, more subtly, for a stake in and commitment to the hegemonic culture of engineering.

The 1976 meeting of SHOT is usually taken as a watershed in the writing of women into the literature of the field. Although Ruth Schwartz Cowan had already presented some of the material that was to find its way into her classic monograph *More Work for Mother: The Ironies of Household Technology from the Open Hearth to the Microwave*, this meeting was the venue for the presentation of her equally classic paper "From Virginia Dare to Virginia Slims: Women and Technology in American Life."[3] In it she famously asked, "Was the female experience of technological change significantly different from the male experience?" She suggested, of course, that the experience *was* different and she suggested four areas where one might look at how: with respect to women's bodies and their functions, in work outside the home, in domestic labor, and in the ideological realm where women were constructed as "not good with machines" and thus cut off from one of the signal marks of full American citizenship. That same bicentennial meeting, moreover, also marked the formation of Women in Technological History (WITH), a special-interest group within SHOT which sought both to ensure that female scholars in the field were fairly represented in the society's leadership and on the annual programs and also to focus attention and research on the relationship of women to technology through time.[4]

In 1979 Martha Moore Trescott, a major figure in the formation of WITH, published a groundbreaking anthology of articles under the title *Dynamos and Virgins Revisited: Women and Technological Change in History.*[5] Much of the research presented in the volume was usefully recuperative, such as Deborah Warner's essay "Women Inventors at the Centennial." Susan J. Kleinberg's essay, "Technology and Women's Work: The Lives of Working Class Women in Pittsburgh, 1870–1900," brilliantly compared the differential impact of industrial technologies and municipal utilities' infrastructure on the work of men and women in and around the steel mills of that city. Not surprisingly, around this time an increasing number of articles and monographs revealed that women had been inventors and engineers, driven cars and flown airplanes, gone to sea in ships, and worked in any number of occupations once thought of exclusively as the purview of men.

Other work looked at occupations traditionally gendered feminine and reassessed them in terms of skill demanded and socially useful work performed. A paradigmatic example was needlework, which had always been assumed to be a feminine occupation and, at the same time, had not been thought of as an occupation at all but more of an "accomplishment" or, at the other extreme, sweated labor. Cooking is another such activity, as indeed is consumption. The recasting of these activities as worthy of study was a powerful result of this first wave of revision of the history of technology canon.

A useful measure of where the fields of the history of technology, of science, and of medicine stood in the late 1970s can be found in the volume *A Guide to the Culture of Science, Technology, and Medicine.*[6] Published in 1980, the volume had been conceived of and jointly sponsored by the National Endowment for the Humanities and the National Science Foundation. Believing, correctly as it turned out, that scholars in the nine fields of the history, sociology, and philosophy of technology, science, and medicine had many overlapping interests and read many of the same books, the two federal agencies recruited an advisory board and authors to represent their fields at that point in time so that the overlaps might be highlighted. The authors were closely monitored and were given a template of topics that were to be addressed by each specialist.

One such common subject was to be areas of particular, concentrated, or emerging interest, termed "Specialized Fields." When the book was published, the index listed only two references to women, both a result of this particular injunction. In his section on the history of medicine, Gert Brieger chose to address the family, childhood and child rearing, women, and the history of education. The section (one page) on women was defined pretty much as "medical care and diseases of women" and carried the caveat: "while the newer work tends to have a broader perspective, some of it has suffered from a feminist ideology based on a conspiracy theory of history."[7] Like many of us, he would probably like to revise that opinion in the light of subsequent work in the field. My own essay on the history of technology also contained a one-page section on women, emphasizing the work of Susan J. Kleinberg and Ruth Cowan from the more recent literature. My remark that "female participation in the making of the technological world has certainly been limited" gives some indication of the distance we have come over the past two decades.[8] The essay on the history of science was silent on the subject of women. All this is indicative, I think, of just how slowly feminist scholarship had begun to change the ways in which these fields were shaped.

In 1982 Judith A. McGaw published a review essay in *Signs* entitled "Women and the History of American Technology," which allowed her to canvass the field as it was emerging.[9] Harking back to Cowan's query of only six years earlier, McGaw was able to point to a growing literature which did, indeed, take seriously the proposition that historically in our society, men and women experienced technology differently. Most of what she called this "prodigious effort in a new and promising field" fell within the four categories outlined by Cowan: the biological (because women "menstruate, parturate and lactate" was the way Cowan phrased it), work outside the home, work inside the home, and the ideological, that is, the perception of women as "antitechnocrats." The first of these, McGaw believed, was the most neglected, although she was able to point to some pioneering work on birth control. The last was also unsatisfactory, and the category itself seemed to make McGaw somewhat uncomfortable, open as it was to too-easy assumptions of essentialism. Studies of women inventors and engineers certainly fell within this category of ideology, but they had been largely recuperative in nature. It was in the categories of work within and without the home that McGaw found the most evidence of scholarly interest.

McGaw's survey of scholarship on women's work outside the home began with Elizabeth Faulkner Baker's classic study *Technology and Woman's Work* (1964),[10] which almost defines the liberal, individualist, recuperative style of women's history. In Baker's story, new technologies opened up new opportunities for women's employment, and increasing numbers of them left the home to work in the larger world. Newer studies, however, like McGaw's own definitive *Most Wonderful Machine: Mechanization and Social Change in Berkshire Paper Making, 1801–1885,* provided a more structured analysis of which technological changes made what differences in which jobs within the mill and of how these jobs were distributed between the sexes. Like other works in this vein, McGaw's found that mechanization was often deployed in such a way as to preserve male privilege and keep female operatives in the lowest paid, often dead-end jobs where their skills were denied.[11]

Certain industries, especially the textile industry, were given more attention than others in which women also labored. McGaw could report on a few attempts to look at the field of office and clerical work as exceptions to this imbalance, and she drew attention to an interest in the way in which wars drew women into manufacturing employment of one kind or another. It was her observation, however, that by and large scholars too often concentrated on wages, hours, and conditions while treating technologies as an unexamined black box.

The study of women's work within the home had been given excitement and legitimacy by Cowan's *More Work for Mother,* but other studies also began to look at the impact of the home economics movement and the spillover from industry of attempts to "rationalize" housework by applying scientific management to domestic chores and arrangements. Since a fair amount of the nation's housework was done by hired domestics, their histories too became a part of this literature, although, as with factory and office workers, technology formed more of a backdrop than a part of the framework to be analyzed. McGaw also sensibly took notice of books on architecture and home building, such as Dolores Hayden's *The Grand Domestic Revolution: A History of Feminist Designs for American Homes, Neighborhoods, and Cities* and Gwendolyn Wright's *Building the Dream: A Social History of Housing in America,* both of which appeared in 1981.[12] Here in the home, as in the factory, technological change appeared to be carefully managed to preserve widely accepted notions of gender-appropriate arrangements and behaviors. Nevertheless, McGaw concluded, "even if historically women have not come a long way, the history of women and American technology certainly has. Its sophistication and scope today are impressive when contrasted with the state of the art only six years ago."[13]

Another six years or so on, McGaw again took the opportunity to assess progress in the field. In a 1989 festschrift for Melvin Kranzberg she published her influential essay "No Passive Victims, No Separate Spheres: A Feminist Perspective on Technology's History." She does not refer to Joan Wallach Scott's now classic article "Gender: A Useful Category of Historical Analysis,"[14] but her point is the same: the liberal, individualist, and recuperative style of feminist history of technology did not go far enough. Gender, as a socially constructed system based on hierarchies of power, must be analyzed to reveal the ways in which *both* men and women are assigned what are considered appropriate roles in the process of technological change. We have, she charges, largely accepted gender stereotypes and, in her phrase, "looked *through* masculine ideology at the past rather than looking *at* masculine ideology in the past."[15] She believes that scholars should abandon the notion that women are essentially passive in their relationship to technology. Another challenge is to view the decisions and activities of men as the work of *men,* and not just people, thus exposing masculinist advantages designed into technologies and technological systems.

From a gender perspective, McGaw could assert that by studying "both the men and the women in industries employing both," we can "begin to see those links that gender rhetoric denied."[16] One insight thus found is that "our sources have . . . given us a nineteenth century 'masculine' definition of production."[17] This was a critical point, and one flowing directly from the analysis of gender as social contract: we were now in a position to see not only that women went whaling, flew airplanes, made cigars, and wove cloth but also that they performed whole categories of "production" that were previously invisible because not seen as such. As she put it, before Levi Strauss produced trousers and Campbell produced soup, women performed the same work in their own homes (or in someone else's), but it somehow was not considered to be production.[18] One of the few trained colonialists among historians of American technology, McGaw was particularly helpful in showing that women did not simply react to the Industrial Revolution but helped create it in the eighteenth and early nineteenth centuries.

The melding of women's history and the history of American technology that began in the 1970s opened up whole new areas for research and greatly enriched the field. The influx of women scholars and women as a topic helped speed the dilution of the old inherited engineering culture. Many members of SHOT continued to come to the field with some engineering background, but increasingly people whose entire training had been historical took positions of influence and leadership: without much discussion, practicing engineers were no longer rotated through the presidency. As the older topics of design and production of large metal technologies lost some of their exclusive claim to legitimacy, alarms were raised that the field was losing its core mission and even its reason for separate existence, but by then enough people were using newer methods and studying newer topics that the skirmishes were increasingly over balance rather than legitimacy.[19]

Eight years after Judith McGaw issued her call for a systematic gendered understanding of the history of American technology, SHOT's journal *Technology and Culture* published its first special issue: *Gender Analysis and the History of Technology*. The guest editors, Nina E. Lerman, Arwen Palmer Mohun, and Ruth Oldenziel, announced on the first page that "gender ideologies play a central role in human interactions with technology, and technology in Western culture is crucial to the ways male and female identities are formed, gender structures defined, and gender ideologies constructed." Their ambitions were appropriately large: "gender analysis of 'ways of making and doing things,'" they asserted, "challenges conventional assumptions about what is and is not 'technology,' and about which technologies are or are not important to study." Furthermore, "many scholars have found that thinking analytically about gender, about maleness and femaleness, leads them to reexamine other standard dichotomies: familiar pairs such as private and public, home and work, consumption and production have come under scrutiny."[20] McGaw had made similar claims earlier in her essay "No Passive Victims," but now a growing number of young scholars were taking this very seriously. And like McGaw on several occasions before them, the editors provided a rich bibliographic survey of the work they found relevant.[21]

Three of the articles in the special gender issue of *Technology and Culture* were contributed by the editors themselves and epitomize the kind of scholarship they both advocated and presented; all three articles were the outgrowths of the authors' dissertations. Lerman's article, " 'Preparing for the Duties and Practical Business of Life': Technological Knowledge and Social Structure in Mid-Nineteenth-Century Philadelphia," shows the ways in which school officials agreed that technical education was important for children but made conscious and deliberate distinctions about race and class as well as gender when assigning "appropriate" skills and technologies to each social group. Oldenziel wrote on the General Motors effort to recruit boys into model-making contests between the wars, and the gendered ways in which that was conceived. Mohun's "Laundrymen Construct Their World: Gender and the Transformation of a Domestic Task to an Industrial Process" treated the rise of steam laundries.[22]

All the authors represented in the special issue set the gendered nature of women's work against the equally gendered nature of men's work. Writing on topics which might well have been tackled a generation ago, these scholars completely reconceptualized them. Instead of ignoring women completely, or merely uncovering the women who did happen to be a part of the story after all, they have told radically different stories of both men and women locked in a reciprocal (though hardly even) struggle over definitions, meanings, and, ultimately, power.[23]

The older style of recuperative women's history is still being written, of course, but having originally inspired this kind of scholarship, feminist theory has moved forward to suggest even more radical and provocative topics and methodologies.[24] Feminist scholarship in the past two decades has made it necessary to include women in our stories, to see white men as privileged and not as universally representative, and to question the subjects and categories we use, and most important of all, it has legitimized—I think even required—a serious consideration of theory as such, especially that arising from the cognate disciplines of literature, anthropology, sociology, and geography.

Perhaps one example will suffice. Historians have in recent years begun to participate in the theorization of race. The debt to feminist theory is clear: just as hegemonic masculinity must be made visible, so too must the construction of racial hierarchies and subordination. Racial characteristics and identities are socially constructed in ways similar to those in which gender is socially constructed. As historians have become aware that we have been telling stories in which women were absent, so have they become aware that we have been telling stories devoid of racial "others." Just as women were deliberately discriminated against in the fields of invention and engineering, so were African Americans and others who were assumed to have "color." To the considerable extent to which closeness to technology was considered a mark of American belonging, the technological disability of both women and racial minorities was a powerful mark of less-than-full participation in the public life of the nation.

But the connections between feminist theory and racial theory are even closer. African American and Hispanic men, for example, are often feminized as a way to stigmatize and marginalize them. As I have attempted to demonstrate elsewhere with respect to the so-called appropriate technologies of the 1970s, machines and processes too can be feminized and therefore stigmatized and marginalized.[25] It has become common to decry the trinity of race, class, and gender, but listen to the concern for the future status of his profession expressed in 1903 by one American engineer: "We have the man who fires the boiler and pulls the throttle dubbed a locomotive or stationary engineer; we have the woman who fires the stove and cooks the dinner dubbed the domestic engineer, and it will not be long before the barefooted African, who pounds the mud into the brick molds, will be calling himself a ceramic engineer."[26] The status anxiety among American engineers revealed here is common enough, even in our own times, but this particular invocation simultaneously of the dangers of race, class, and gender in an unconscious definition of what, to his mind, engineering was not—and therefore was—suggests to me the ease with which these categories can be combined or substituted for each other. Feminist theory has sensitized us to such language and such strategies, and it has suggested ways in which these can be investigated and understood.

Just as an awareness of gender reinforces an awareness of race, so does the invocation of race remind us that any discussion of women and technology that ignores the issue of race is incomplete. Like gender, race in the United States saturates every aspect of our lives and thinking. Just as the dearth of women scholars in the history of technology in its early years helped masculinist values masquerade as universal, so has our lack of racial diversity allowed us to assume that whiteness is normal. Even the growing number of studies of women and technology virtually ignores issues of race. Lerman and Mohun both confront race directly in their dissertations and publications, but we need much more of this. Bruce Sinclair has undertaken a project to produce a two-volume study that will provide theoretical and hortatory articles on the subject of African Americans and technology, scholarly examples of what such work might look like, and a collection of documents as well.[27] If we are to take whiteness seriously as a racial construct, we must also begin to look at how racial privilege and technological privilege have been mutually constitutive.

It is my own strong belief that it is time to press on now and extend the feminist project in several directions: it has evolved wonderfully over the past two decades and will continue to do so. One direction is toward a growing embrace of theory. One of the most powerful features of feminist theory is that it speaks across disciplinary lines and across the chasms of specialization that too often divide even historians of technology from one another. Whatever one's period or geographical focus, whether one works on steel or textiles, design or repair, feminist theory is a necessary tool of analysis, and one that can be shared and appreciated by colleagues. In a similar way it provides a bridge between ourselves and colleagues situated in other disciplines, who theorize gender and sexuality as performance (might technology, in some sense, be performative as well?), map gendered spaces in cities (what part does technology play in this?), or explore through cyborgs the cultural relationships between the body and machines.

Another direction is to use feminist theory to suggest ways in which we can continue to cast off inherited categories and hierarchies that are constrained by race and class as well as by gender.[28] Race and class are categories that include women as well as men and that affect the experience of gender in ways which are at once familiar and different. Until these too are part of our stories, and their constructedness and operations a part of our understanding of technology and how it operates, our feminist project remains only partially fulfilled. Although class as a social category features in a great library of historical studies, it comes close to technology only as labor history, and that only on occasion. The class bias in considering only design as an important technological activity should, by now, be obvious.

If the influence of feminism on the scholarly study of the history of technology is clear enough, its influence on technology itself is less so. In the mid-1970s, when feminists and others were urging young women to enter the engineering profession, this question was often posed as, Will the presence of women engineers change engineering, or will engineering change the women who take it up? The same might be asked of any of the design professions—invention and architecture, for example. It must be realized, however, that the very question smacks of a dangerous essentialism—as though, because women were *by nature* more gentle and morally superior, anything they might design would have these same characteristics. My own feminism leads me to reject such essentialist thinking, but, of course, this still leaves open the very real possibility that women might be tracked into certain branches of these fields. It also seems highly possible (and hardly surprising) that many women are acculturated to accept and some have even internalized what seem "appropriately feminine" behavior and expectations.

It has been noted, of both women and men inventors, that individuals tend to invent most successfully in those areas they know best. People close to agricultural tools and practice disproportionately invent agricultural implements; people in the armament industries are most likely to invent new weapons. It would not be surprising, therefore, if women inventors worked most often with appliances most closely associated with those areas in which women have been most numerous and approved, for example, housework and fashion and child care. Madame C. J. Walker, one of the best-known African American inventors, became an early millionaire through the invention and marketing of hair products for African American women.[29] More recently, it was a woman, Hazel Bishop, who invented nonsmear ("kissproof") lipstick.[30] If these examples seem trivial, it is perhaps in part because a gendered culture tends to trivialize any things or activities culturally associated with women.

Women in architecture might be similarly tracked. Years ago an architect of my acquaintance made a name for herself redesigning whole floors of hospitals into more patient- and family-friendly birthing suites. Dolores Hayden has written extensively about feminist architects who chose to concentrate on dwellings, in an effort to design into them their own concepts of how women (and men) should live. The kitchenless house is perhaps the best-remembered example.[31]

And it was the kitchen that provided a career for the psychologist-turned-engineer Lillian Gilbreth after her husband and partner died.[32] Clients turned away from her on the assumption that a woman could not know industrial sites and processes, but she was famously able to exploit gender expectations to champion the "Kitchen Efficient." It is perhaps ironic that Gilbreth should have been the model figure for the Society of Women Engineers, founded in 1950. From the beginning the new organization found itself trapped between two missions: attracting more women into the profession and giving badly needed support for those already at work as engineers. The first, of course, led the organization to downplay the gender bias and misogyny that permeated the field; the second required that the problems be admitted, faced, and worked against. One possible problem with the notion that women might actually change engineering, in this case the areas in which research was undertaken, was raised by Dr. Maria Telkes, the first recipient of the Achievement Award given by SWE in 1952. Her work was on solar energy, which over the years had been recognized and funded less than nuclear energy research because, as she put it, "sunshine isn't lethal."[33] Whatever her choice of research field, Telkes did her work within a larger framework of rewards over which she had very little influence. Every technological design is encoded with human intention, however, and it is this characteristic which makes possible the social and cultural "construction" of technologies in ways which can be seen as oppositional.

But to limit one's consideration to the impact of feminism on the traditionally masculine design fields such as invention and engineering is to substitute a part for the whole. One might more fruitfully look at the subject of women's entry into the construction trades or any of the other technical areas of work that have been retained and policed by powerful male unions. The organization Hard-Hatted Women, for example, promotes and supports women in the building trades in ways that are arguably as successful, and perhaps more overtly feminist, than SWE in engineering.[34]

One final place in which to search for feminist influences might be the area that Ruth Schwartz Cowan called "ideological" in 1979, that is, women as "antitechnocrats." As she noted, "if practicality and know-how and willingness to get your hands dirty down there with the least of them are signatures of the true American, then we have been systematically training more than half of our population to be un-American. I speak, of course, of women."[35] She carefully uses the term "antitechnocrats," not "antitechnology," which I take to be a suggestion that the women she has in mind reject, not technology as such, but the masculinist forms in which it is so often delivered and celebrated. Carolyn Merchant's essay "The Women in the Progressive Conservation Crusade: 1900–1915" highlights the fact that these women "frequently saw themselves as ideologically opposed to what they perceived as commercial and material values." They were, she concluded, "feminist and progressive in their role as activists for the public interest."[36] The high proportion of women in positions of environmental authority nearly a century later can perhaps be traced to the same causes.

The Appropriate Technology (AT) movement of the 1970s was also feminized, but in this case by opponents who were wedded to either the culture or the political economy of large technological systems. Technologies such as those providing solar and wind energy, sustainable agriculture, recycling, and mass transit were painted as both un-American and unmanly.[37] There is evidence that some feminists were attracted to the opportunities the AT movement seemed to offer to rethink the relationships between women and technology, but there is equal evidence that the reality of gender relations within the movement fell far short of their ideals. In 1979, for example, two women from the Santa Clara County (California) Office of Appropriate Technology organized a series of workshop/study groups under the rubric "Women and Technology." "Through our involvement in appropriate technology," they wrote, "and our feminist orientation, we've become concerned with the connections between the two and we need to integrate them in our work."[38] At the same time, others noted that there was nothing necessarily feminist about AT as such. Judy Smith, who worked at the Missoula, Montana, headquarters of the National Center for Appropriate Technology, charged that in writing a pamphlet entitled *Something Old, Something New, Something Borrowed, Something Due: Women and Appropriate Technology,* she "found that people who espoused those ideas [of AT] and who led the movement were men, making decisions based on the same old values." She concluded that "sex role separation continues even within these 'advanced' movements." The "most important task ahead," she claimed, "is involving women in technological decision making."[39] Clearly what attracted these women to the AT movement was the implied opportunity for them to help redesign the technological systems with which they lived, not the specifics of any one technology that might be handed to them.

The question of how much and in what ways feminists have managed to shape our technological regime in twentieth-century America is a fascinating one and promises, when seriously addressed, to add to both our academic understanding of how technologies change and to our political power to affect such changes. The changes wrought thus far by feminist scholarship in the field of the history of technology have already shown the way. By deprivileging those categories of inventor and engineer which white men have largely kept to themselves, we can begin to better see the women as well as the men, the blue-collar as well as the white-collar workers, and the racialized "people of color"—of whatever sex—who have always dealt with the full range of technologies throughout American history.

NOTES

1. Judith A. McGaw, "Women and the History of American Technology," *Signs: Journal of Women in Culture and Society* 7 (1982): 798–828.

2. Judith A. McGaw, "No Passive Victims, No Separate Spheres: A Feminist Perspective on Technology's History," in *In Context: History and the History of Technology—Essays in Honor of Melvin Kranzberg,* ed. Stephen H. Cutcliffe and Robert C. Post (Bethlehem, PA: Lehigh University Press, 1989), 172–91.

3. Ruth Schwartz Cowan, *More Work for Mother: The Ironies of Household Technology from the Open Hearth to the Microwave* (New York: Basic Books, 1983); and Ruth Schwartz Cowan, "From Virginia Dare to Virginia Slims: Women and Technology in American Life," *Technology and Culture* 20 (Jan. 1979): 51–63.

4. *Women in Technological History: A Look Back* (n.p.: WITH, 1997).

5. Martha Moore Trescott, ed., *Dynamos and Virgins Revisited: Women and Technological Change in History* (Metuchen, NJ: Scarecrow Press, 1979).

6. Paul Durbin, ed., *A Guide to the Culture of Science, Technology, and Medicine* (New York: Free Press, 1980).

7. Ibid., 153.

8. Ibid., 88–89.

9. McGaw, "Women and the History of American Technology."

10. Elizabeth Faulkner Baker, *Technology and Woman's Work* (New York: Columbia University Press, 1964).

11. Judith A. McGaw, *Most Wonderful Machine: Mechanization and Social Change in Berkshire Paper Making, 1801–1885* (Princeton, NJ: Princeton University Press, 1987).

12. Dolores Hayden, *The Grand Domestic Revolution: A History of Feminist Designs for American Homes, Neighborhoods, and Cities* (Cambridge, MA: MIT Press, 1981); and Gwendolyn Wright, *Building the Dream: A Social History of Housing in America* (New York: Pantheon Books, 1981).

13. McGaw, "Women and the History of American Technology," 828.

14. Joan Wallach Scott, "Gender: A Useful Category of Historical Analysis," in her *Gender and the Politics of History* (New York: Columbia University Press, 1988), 28–50; first published in *American Historical Review* 91 (1986): 1053–75.

15. McGaw, "No Passive Victims," 177.

16. Ibid., 178.

17. Ibid., 179.

18. Ibid.

19. See Rosalind Williams's useful intervention in the debate over "technological determinism": "The Political and Feminist Dimensions of Technological De-

terminism," in *Does Technology Drive History? The Dilemma of Technological Determinism,* ed. Merritt Roe Smith and Leo Marx (Cambridge, MA: MIT Press, 1994), 217–35.

20. Nina E. Lerman, Arwen Palmer Mohun, and Ruth Oldenziel, "Versatile Tools: Gender Analysis and the History of Technology," in *Gender Analysis and the History of Technology,* ed. Nina E. Lerman, Arwen Palmer Mohun, and Ruth Oldenziel, special issue of *Technology and Culture* 38 (1997): 1, 3, 5–6. Their use of the terms "maleness" and "femaleness" borders, however, on essentialism: one wonders why the terms "masculinities" and "femininities" were not used instead.

21. Nina E. Lerman, Arwen Palmer Mohun, and Ruth Oldenziel, "The Shoulders We Stand on and the View from Here: Historiography and Directions for Research," *Technology and Culture* 38 (1997): 9–30.

22. Arwen P. Mohun, *Steam Laundries: Gender, Technology, and Work in the United States and Great Britain, 1880–1940* (Baltimore, MD: Johns Hopkins University Press, 1999).

23. Carroll Pursell, "The Construction of Masculinity and Technology," *Polhem* 11 (1993): 206–19; and Judy Wajcman, *Feminism Confronts Technology* (University Park: Pennsylvania State University Press, 1991).

24. See the review essay by Judith McGaw, "Inventors and Other Great Women: Toward a Feminist History of Technological Luminaries," *Technology and Culture* 38 (1997): 214–31.

25. Carroll Pursell, "The Rise and Fall of the Appropriate Technology Movement in the United States, 1965–1985," *Technology and Culture* 34 (1993): 629–37.

26. Quoted in Bruce E. Seely, "SHOT, the History of Technology, and Engineering Education," *Technology and Culture* 36 (1995): 744.

27. This work is being supported by the Lemelson Center at the Smithsonian Institution.

28. See my article, "Seeing the Invisible: New Perspectives in the History of Technology," *Icon* 1 (1995): 9–15.

29. Portia P. James, *The Real McCoy: African-American Invention and Innovation, 1619–1930* (Washington, DC: Smithsonian Institution Press, 1989), 85–86.

30. "Hazel Bishop, 92, an Innovator Who Made Lipstick Kissproof," obituary, *New York Times,* Dec. 10, 1998, sec. B, p. 16, col. 3.

31. Hayden, *The Grand Domestic Revolution.*

32. On the general subject, see Ruth Oldenziel, "Decoding the Silence: Women Engineers and Male Culture in the U.S., 1878–1951," *History and Technology* 14 (1997): 65–95.

33. Society of Women Engineers, *Achievement Awards, 1952–1974* (n.p.: SWE, n.d.).

34. For a collection of first-person accounts, see Molly Martin, ed., *Hard-Hatted Women: Stories of Struggle and Success in the Trades* (Seattle: Seal Press, 1988).

35. Cowan, "From Virginia Dare to Virginia Slims," 61–62.

36. Carolyn Merchant, "The Women in the Progressive Conservation Crusade: 1900–1915," in *Environmental History: Critical Issues in Comparative Perspective,* ed. Kendall E. Bailes (Lanham, MD: University Press of America, 1985), 170.

37. See Pursell, "The Rise and Fall of the Appropriate Technology Movement."

38. Circular signed by Carol Manahan and Gina Moreland, dated June 12, 1979.

39. Judy Smith, "Women and Appropriate Technology: A Feminist Assessment," in *The Technological Woman: Interfacing with Tomorrow,* ed. Jan Zimmerman (New York: Praeger, 1983), 66, 69.

SEVEN

Man the Maker, Woman the Consumer: The Consumption Junction Revisited

RUTH OLDENZIEL

What counts as technology or who is to be considered a technologist goes to the heart of contemporary feminist inquiry. The often posed question of why women have failed to enter the male technical domain puts the blame entirely on women: it blames them for their supposedly inadequate socialization, lack of aspiration, and want of masculine values. An exclusive focus on women's supposed failure to enter the field of engineering, however, does not explain how our stereotypical notions of men as active producers or designers and women as passive consumers or users have come into being. Our contemporary gender mythologies imply that women must be amateurs suffering from stage fright whenever they enter the male constructed stage of technology or, by the same reverse logic, that technology is what men do and women do not.[1] At best, our modern myths of technology frame a world in which men design systems and women use them, men build bridges and women cross them, men design cars and women ride in them; in short, a world in which men are considered the active producers and women the passive consumers of technology. Man the maker and woman the consumer are two sides of the same coin circulating in our ideological economy. Technology has become one of the most important pillars of modern manliness.[2]

Our challenge is to understand how our contemporary mythologies have come to designate the things men do as technology and women's activities as nontechnical; how in the twentieth century, technology has been transformed into a powerful symbol of modern manliness that is nevertheless socially constructed and historically contingent; and how man the maker and woman the consumer have mutually shaped constructions of gender and technology.

These questions are both historiographical and historical in nature. In the pages that follow, I will therefore briefly explain the historiographical hurdles which still prevent historians of technology from seeing women as active agents in technological developments and from understanding gender as a powerful category of analysis. I then turn to historical examples in which, contrary to our contemporary mythologies, women from both inside and outside the women's movement appear as active agents in the shaping of technology in the last hundred years or so. Finally, I return to the implications these historical examples have for the kind of questions we might ask. They give a clue to the most promising lines of research when trying to bring into sharper focus the issue of women's agency in our histories of technology.

Despite critical assessments of feminist scholars over the last two decades, most scholars continue to have a hard time perceiving women as active agents in the history of technology or understanding gender formation as part of technological developments. Framed as they are by the history of engineering and science, even the most ambitious and progressive histories of technology have been hampered rather than helped when selecting concepts, primary sources, and periodization. Many historians of technology—among them American, German, and Dutch researchers of the earlier generation—were either lapsed engineers who sought to upgrade their profession by mobilizing history or historians teaching at technical institutions who were looking for ways to make their teaching more meaningful for their students.[3] They canonized engineers and inventors as the true bearers of technology. This engineering genesis of the field unwittingly excluded those historical actors operating outside its definitions of the engineering professions.[4] Through these narrowly focused lenses, even the few women engineers, inventors, and architects who succeeded appear to be of no consequence.

The engineering frame in the history of technology often goes hand in hand with an emphasis on business history, reinforcing what one could call a paradigm of production. It places great emphasis on tangible products, objects, hard work, and character rather than on knowledge, institutions, users, leisure, and personality.[5] In this frame, technical artifacts guide decisions for case studies, periodization, and definitions. More specifically, it privileges design over use, patent activity over tacit knowledge, engineering products over nonengineering products, capital-intensive technologies over labor-intensive technologies, and—as a consequence—white men over women, workers, and ethnic and non-Western people. Even in such pioneering sociological studies by Cockburn and Ormrod on the microwave oven and the role gender has played in its shaping, the main emphasis remains on design regimes, production processes, and projected users rather than on the extended communities of repair, maintenance, and use.[6]

The third corollary of the historiographical focus on production is generated by its implied opposite. The world of production supposes a world of consumption where users eagerly and passively await producers' merchandise. In this world, women are technology's passive consumers and end users only. Readers of gender studies may easily recognize this bias as yet another incarnation of the Victorian and middle-class ideology of the separate spheres where middle-class men tend to the coldhearted industrial world of work and their female counterparts are left to cultivate the world of domesticity in order to offset the suffering and ruthless competition of the capitalist system.[7] These constructed dichotomies have found their way into most histories of technology as well.

RESISTING MAN THE MAKER AND WOMAN THE CONSUMER, 1870–1915

The modern notion that men make things, design bridges, and engineer our technologies—and women use them—represents a material, political, and cultural order bound by time and place. These dichotomies are part of industrial capitalism's narrative productions. The nineteenth-century mobilization of cultural resources for the new industrial world involved narrative productions that included popular literature, advertising copy, encyclopedias, metaphors, and keywords of America's modernist grammar.[8] Taking hardware—machines in particular—as the exclusive site of technology has become dominant only since World War I, after a century-long contest over technology's meanings. In the nineteenth century, for example, inventions ranged from corsets and bonnets through arts, music, and language to the crafts. A hundred years later, all these products had been banished to the basement of modern classification. Instead, the material products of (mechanical and civil) engineering became the markers of true technology. In this new productionist paradigm of technology, machines became the metaphor, model, and microcosm of technology—and, in this disguise, of male Western power, too. To be sure, the focus on, fascination with, and fetishization of commodities were part of a certain stage of industrial capitalism, in which machines were both embodiments of products as well as the producers.[9]

Thus, machines became the true fetishes of modernity. Middle-class men were taught to see their manliness reflected in the clean surfaces of streamlined machines.[10] In America and elsewhere in the Western world, the building of the institutional, cultural, and social infrastructures that began to buttress this view of technology did not come about without the protest of women. Throughout the nineteenth century, activists made themselves heard at various important staging grounds of industrial capitalism, ranging from the hall of the Patent Office to fairgrounds.

Women thinkers and writers from the radical Joslyn Gage to the conservative Ida M. Tarbell wrestled and debated with the emerging genealogy that began to reinscribe inventions—mechanical ones in particular—as male. In the 1870s, the women's rights activist Gage (1826–98) argued that the patent system protecting intellectual property could never be fair if women had no political clout or could not legally own the patents they submitted. She believed that conservative women's efforts to promote women's inventions were doomed from the start. In the 1880s, the journalist, business historian, and lecturer Tarbell (1857–1944) argued that the problem of women's low invention rate was due to the devaluation of the kind of inventions and skills in which women excelled, such as needlework. Tarbell, who would later expose the monopoly practices of Standard Oil, believed that the emerging taxonomy unjustly privileged such male-centered technologies as the machine-tools industry and mechanical engineering over such female-centered ones as needlework.

If Gage, Susan B. Anthony, Elizabeth Cady Stanton, and others engaged in confrontational tactics to protest exclusionary politics, other women's groups also negotiated the material practices and intellectual constructs of the new capitalist order that began to privilege capital-intensive over labor-intensive technologies.[11] In temporary coalitions, individual entrepreneurial women and women's rights activists demanded the rightful place and fair treatment of the hundreds of women inventors. Widows and working women with small incomes trying to cash in on their inventions joined hands with women's rights activists interested in mobilizing exemplars that could demonstrate women were men's equals in the progress of civilization and should therefore be treated as full citizens of the polity. At one of the new temples of industrial capitalism, the Patent Office in Washington, D.C., they staged a protest for fair treatment in 1890.

Middle-class women, organized into lobby groups, contested the emerging male genealogy also on the grounds of world's fairs, where nations competed with each other by linking patent activity to national progress. World's fairs helped to shape the institutional, cultural, and social infrastructures that came to buttress the new male authority of science and technology. Nineteenth-century world's fairs—true carnivals of industrial capitalism—were the contested terrains of the new classification of the industrial world where these classifications served material, visible, and symbolic roles, as many women's advocates understood so well. Sensing the importance of these classificatory frameworks, middle-class women activists began to challenge them; they invested enormous financial, organizational, and intellectual resources in several world's fairs in the United States and beyond, in an effort to counter the fairs' growing importance as new sources of male authority. Both moderate and radical women's activists petitioned for congressional funds, staged protests, planned sessions, organized alternative programs, and mounted exhibits to display women's skills, products, and inventions at a number of world's fairs. They displayed women's skills at fairgrounds in Philadelphia (1876), Nashville (1884), Chicago (1892), Atlanta (1895), Charleston (1901), St. Louis (1904), Portland (1905), and San Francisco (1915).[12] They sought to correct the male genealogies, fully aware that through deed and word the emerging male taxonomy of the material world of industrial capitalism neglected those groups which had been excluded from the official institutions of invention, engineering, and science. These examples illustrate that throughout the nineteenth century women were active technical agents, and not passive bystanders, in the shaping of one of the most prominent symbols of modern manliness: the world of inventions, machines, and production. Up until World War I, women continued to lobby for the inclusion of women's inventive genius and to protest against the male rewriting of the history of inventions as their exclusive domain and the badge of modern manliness. Nevertheless, after the war the discourse on middle-class men as producers of technology became dominant. Professionalizing engineers and their advocates, together with the new powerful corporations, began to represent themselves as the sole bearers and producers of technology. Women's activities in bonnet-making, sewing, bread-making, and the like were now relegated to the basement of modern classification systems as true technologies. A paradigm of production had been born. So, too, its polar opposite.

"It is the extent of obstinacy or the disinclination of women to use labor-saving inventions, . . . which deserves, if anything does, to be called feminine and to be utilized as a warrant for masculine vaunting of superiority in this particular direction," one reader claimed in the *New York Times* of 1915. In the emerging modernist discourse of the twentieth century, women's perceived reluctance to buy laborsaving devices was increasingly castigated as irrational and dismissed as feminine, whereas men's love for things technical increasingly defined their manliness and superiority. On the other side of the Atlantic similar casually uttered remarks were heard, casting women as hopelessly antimodern, irrational, and technophobic. Women replied with equal vigor to these comments. In a letter to the editor, Mary Ritter Beard (1876–1958), historian, author, and suffragist, argued that men's power of the purse was to be blamed rather than women's resistance to new technologies. Farm women, for example, reported to the federal Department of Agriculture that "while their husbands have the family cash with which to buy reapers and other devices, the wives are given no control over the family earnings, which would enable them to buy household devices to lighten their own labor." She hoped many other utility companies and enlightened city officials would follow the example of one Progressive town that supplied cheap electricity to all homes. She expected that once women had the vote the new technological development in electricity would be directed toward relieving the burden of women's household work. "Votes for women might very well lead to a great inventive stimulus and the means of the possible application of domestic appliances to household drudgery."[13] During her lifetime, she witnessed and understood the making of a modernist discourse that increasingly rendered women as the passive bystanders of history. From her early book *Woman's Work in Municipalities* (1915) to her most famous book, *Woman as Force in History* (1946), Beard sought to set the historical record straight by arguing that women were a powerful force in the creation of civilization and culture and that a history that did not include them would be incomplete.

The engineering-productionist paradigm of the history of technology that Beard challenged was part of the larger masculinist narrative of industrial capitalism. As more and more women entered the labor market and new professional domains, male authority was revitalized through the culture of production that insisted, "categorically, man is the producer, woman the consumer," as *McCall's* echoed the cliché in 1937.[14] It began freezing out entire fields in which women dominated—such as home economics, bread-making, and sewing—as legitimate technological subjects. It also denied the role consumers played in technological developments. The making of a productionist paradigm involved a modernist-corporate discourse on efficiency, machine aesthetics, and laborsaving that must be carefully taken apart, as Beard's letter to the editor reminds us. The discourse encoded in business records, advertisements, and professional journals, and maintained by the community of producers, needs to be checked against that of the users and the maintenance, service, and repair community rather than simply reproduced. It is not only a matter of "fair play." Using a different framing device also provides insight into the subtle negotiations between producers and consumers in the emerging consumer society of the twentieth century.

To explore anew the formation of consumer culture in the twentieth century from the point of view of the history of technology is to see how new institutions shaped the communities of producers and consumers and to challenge the exclusive construction of technical artifacts as being solely the work of inventors and manufacturers. Such an exploration shows the integral role women have played in the technical developments. It restores them to their rightful place as active producers of technical artifacts rather than as passive bystanders.

REVISITING TECHNOLOGY'S CONSUMPTION JUNCTION

As Ruth Schwartz Cowan has theorized, the trajectory of a new technology should be mapped from a user's point of view. By way of example, she challenged the notion that Americans, and by implication American housewives, were slow in adopting technically superior and efficient stoves and washing machines, clinging instead to the open hearth and wringers. The focus on the community of engineers, designers, and industrialists had prevented historians from offering any satisfactory explanation; all they did was put the blame on housewives as irrational consumers.[15] Joy Parr has since elaborated that male stove designers and salesmen created products and sales campaigns that clashed with the food preparation practices of the women to whom they were trying to sell. So, too, women's reluctance to adopt washing machines was part of a sensible decision-making strategy. To the hardworking women—who were steeped in a tradition of thrift, prudence, and conservation—these machines were too expensive, complicated, wasteful, and difficult to repair. For them, wringers were the simple, durable, inexpensive, and easy-to-repair washing tools they preferred over unreliable and expensive washing machines.[16]

Users might have existed as configured consumers in the designers' prototypes but they also had a mind of their own. During the 1910s, American farm men and women tinkered with their Model T's and suggested design improvements which Ford incorporated in later models.[17] Some tinkering consumers went so far as to change the intended use and meaning of the car altogether: farm women converted automobile engines into generators for their refrigerators. These subtle subversions, alterations, and modifications and the outright refusal of consumers were all part of the mutual shaping of the design and adoption process between the communities of design and production and that of use, repair, maintenance, and recycling.[18]

Thus users and consumers are neither obedient actors nor passive victims but active participants in shaping technologies. For one, consumers produce frames of meaning that are always part of a technology's ultimately successful adoption. A technology can exist only if people can use it repeatedly, successfully, and daily, as the housewives who preferred wringers over washing machines understood so well.[19]

Technologies—perhaps more so than scientific practices—are to a large degree constituted through their use. A technology, then, is not technical but essentially social. Developing countries seeking to import Western technologies and Western companies interested in selling their products have long understood this truism: technologies and societies need to "fit" for technology transfer to be successful. Through trial and error, businesses have learned that they need to invest an increasingly larger amount in sales departments and advertisement agencies than in the actual manufacturing of their products. This is even true for military technologies, which depend on the mobilization of political legitimacy rather than on commercial success. With their large financial investments and political lobbies, companies in effect seek to produce a "proper frame of meaning" for their product so that potential users will be willing or able to integrate the new artifact into their daily work routines and organizations. Particularly when technologies have not yet found their proper cultural "script," there is more room for negotiation for users as individual tinkerers or as organized groups.

Cultural repertoires are thus always essential building blocks in the making of technologies. Inventors "invent both artifacts and frames of meaning that guide how they manufacture and market their creations."[20] Inventors, despite their rhetoric, do not just "solve" problems, they "frame" them. In the world of invention, problems never exist in a vacuum but are always framed by the solutions inventors like to offer with their ingenuity. Inventors invent problems as well as their solutions. In the 1890s, Thomas Edison, the otherwise successful inventor-entrepreneur, failed to develop the kinetoscope into a successful product because he felt most comfortable in the world of production he understood best: he therefore framed his invention as a tool that would solve a business problem rather than as something that would provide entertainment. In his design, Edison configured the potential user as a white male businessman rather than as a woman or a worker from a different ethnic background.

In the 1890s, Edison was not alone in his miscalculation. In their projections of new technologies, the inventor-entrepreneurs of Edison's generation were plagued by a male productionist frame in the same way that scholars of technology have been in their narrations of technology. As industrial capitalism went into overdrive at the end of the nineteenth century, a whole generation of products were designed initially with male users in mind. Records were marketed as spoken letters and as a form of documentation for the office. Telegraph men designed the early telephones as a means of business communication rather than of social intercourse. In the 1880s, mechanics developed the first bicycles as macho machines with an emphasis on speed, danger, and adventure that catered to the tastes and lifestyles of younger urban white-collar men rather than women and the elderly.[21] In their early designs of the 1910s, radio manufacturers targeted, not listening families in the parlor, but those male tinkerers who devoted all their time to tuning the components to receive radio waves.[22] So, too, early combustion cars of this era targeted upper-class men and male farmers rather than upper- and middle-class women.[23]

What businessmen and historians of technology alike had to recognize is that the new products of the twentieth century—including records, telephones, bicycles, and cars—became successful technologies only when women were included in the design configuration. In many cases, the early inventor-entrepreneurs accidentally "discovered" a new mass market of women. Only when the Parisian makers of carriages for children and the disabled introduced a new bicycle design based on safety did the macho bicycle become a middle-class mass product that would appeal to women and older men. In the 1890s, the safety bicycle—with its equal-sized wheels and inflated tires—enabled a previously unavailable type of travel.

The telephone, too, became a real financial success after businessmen had learned to appreciate the commercial value of women's telephone conversations in maintaining their social relations.[24] The early generation of telephone and telegraph manufacturers had considered women's residential use of the telephone as improper; the users they had in mind were businessmen like themselves. They pushed the telephone as a utilitarian tool for short, useful, efficient, and businesslike conversations and viewed women's telephone conversations as gossip, as a waste of time, and even as acts of subversion. After their initial attempts to suppress women's "improper" use of the telephone, the companies changed their marketing strategy and began to exploit and even encourage and expand on women's telephone habits as a tool for sociability, or, as the ads now insist, "to reach out and touch someone." Having thus reframed the meaning of the telephone, communication companies no longer charged for making the connection but—more profitably—according to how long the call lasted.

These early technological developments of consumer goods demonstrate that entrepreneurs first discovered the female market by accident; their initiatives were often fraught with errors and false starts. This changed as soon as the business community and women's organizations sought to shape the female market. Women as individuals and those organized into groups were part and parcel of this newly created landscape, helping to initiate or frame new meanings for new products in some cases and rejecting products in others. Women were active both inside and outside the women's movement in shaping technological developments. As initiators, designers, lobbyists, and consumer activists they became coproducers of some important technical systems of the twentieth century: city infrastructures, electrical systems, and product engineering.

WOMEN CONSTRUCTING TECHNOLOGY, 1890s–1940s

Businesses, in coalition with women's consumer groups, activists, and home economics professionals, began building a new infrastructure that both discovered and constructed women as consumers of the new technologies. In the twentieth century, the realms of production and consumption increasingly separated what once had been in the hands of members of households. The chains between production and consumption became longer, more diffuse, and less visible. To close this gap, a multitude of mediating institutions, professional groups, and new disciplines developed. These included consumer and market research departments, advertising agencies, research and development divisions, testing places, industrial design departments, housekeeping journals, consumer organizations, home economics professionals, and women's clubs. They were the mediators or translators between the desire of manufacturers to sell products and the necessity of buyers, increasingly women, to incorporate them into their daily practices and work routines.

Around 1900, the women's movement articulated two ideologies that empowered middle-class women to be coproducers of technology. The first was the powerful notion that women as consumers had a role to play in the political economy. The second was the equally powerful idea that women were municipal housekeepers of the industrializing world. From the 1890s until the 1940s, these related notions created a powerful framework for a distinct women's culture of technology.

The National Consumers League—a virtually all-women lobby group—was the earliest organization to argue for the economic importance of consumers. It challenged private industry on behalf of consumers. In doing so, it put middle-class women squarely at the consumption junction of technological developments. Denied the right to vote, the league's middle-class women—as critical consumers—exercised their political power outside the regular political channels through the power of the purse. Under the inspiring leadership of Florence Kelley, they constructed a community where producers and consumers believed themselves to be mutually responsible. In this manner, the league was the first critically important consumer movement, as recognized early on by economic theorists, including the dean of economics Alfred Marshall. Through the White Label campaign, Kelley and other activists sought to battle the exploitative conditions under which a large number of women garment-workers labored to manufacture the kind of consumer goods middle-class women bought. Manufacturers would be allowed to carry the league's white label as a stamp of approval only if certain conditions were met during the manufacture of their products. Based on a network of grassroots organizations that lobbied city councils and state legislatures, the league turned middle-class women into activists as well as into knowledgeable, critical, and activist consumers. This resulted in a new understanding of the consumer's economic centrality—in the new critical knowledge about specific conditions within the local situation, and the activists' newly acquired ability to put that knowledge to political use and to implement legislation. The league's activism began riding the wave of Progressive reform in the 1910s and was its most successful expression of women's political activism.[25] Understanding the new power of middle-class consumers, some firms (including John Wanamaker's department store) engineered a coalition with the league to their own benefit. Through its lobby work, the league effectively shaped the manufacturing technologies and working conditions under which garments were made.

Women's ability to shape technologies became more distinctive in the Progressive Era. The women who participated in the reform movement of the 1910s helped shape an alternative women's technical culture that was nurtured by women's traditions. Until the Progressive Era, the American suffrage movement had based its arguments on appeals to justice and equality. It had drawn its support almost exclusively from middle-class women, but its social and ideological ground was significantly broadened at the beginning of the twentieth century with the formulation of the notion of municipal housekeeping. Middle-class women's domesticity was no longer limited to the physical four walls of the home but extended to the city at large. The reformulation of the home helped to open avenues to careers for women in academia and social reform and also helped them to venture out into the industrial world and to shape it. Under the umbrella of this new ideology, over a million women organized themselves into a large reform movement that cut across different classes and constituencies. Through such pioneering organizations as the Young Women's Christian Association, the Women's Trade Union League, and the Settlement House Movement, they pushed a broad coalition of reform on issues related to women and children. United in the General Federation of Women's Clubs, women were building their own structures. Home economists of Ellen Richard's generation formed an important component of women's political culture during the Progressive Era. The early home economics professionals in particular played a crucial role in politicizing the cult of domesticity, transforming it into a vehicle with which to expand women's political power, and enabling them to enter the public domain as career women or as active agents in public affairs. Under the camouflage of domesticity, the early generation of home economists in particular reinscribed the discourse of the home as a mechanism of female empowerment.[26]

From Boston to San Francisco, women reformers helped build the public infrastructures in the civic improvements movement as private citizens rather than as corporate employees.[27] Highly organized into private philanthropic organizations, like the General Federation of Women's Clubs, these women reformers effectively became the "municipal housekeepers" of the world. They conducted surveys, drew up plans for urban infrastructures, pushed for better housing, and helped finance public facilities, ranging from streetlights to sewer systems. They forged coalitions with local politicians, architects, civic leaders, and professional women like Ellen Swallow Richards, Alice Hamilton, and Ruth Carson in public health, science, and social research. As historian and suffragist Mary Beard demonstrated in her *Woman's Work in Municipalities,* published in 1915, these women of the Progressive civic movement became the urban planners of the modern age.[28]

Incidentally, this kind of distinct female building tradition fell outside the early modern discourse of engineering and technology that I discussed above. The Progressive women reformers were nurtured and sustained by their ties to the women's movement and operated outside the military-industrial complex and patriarchal family firms. In contrast, American women engineers worked within these government and corporate bureaucracies. They relied on ideologies of professionalism, on the mantle of science and engineering, and on the government's recruiting efforts among women to compensate for the country's labor shortage. In fact, the women who belonged to these separate technical cultures—one public and one private—rarely met or built bridges between themselves.[29]

After World War I, the move of women professionals into corporate and government offices both weakened and strengthened women's individual and collective ability to shape technological developments. Many of the postsuffrage generation moved away from a Progressive ethos of reform and municipal housekeeping. They lost their sense of a social and women-oriented mission. Instead, many college-educated women of the next generation moved into corporate and federal bureaucracies. This was due not so much to a loss of gender identity as to the success of the women's movement. Women no longer operated outside social and political institutions but moved into and transformed them, even though the institutions were sometimes beyond their control.[30] Particularly during the 1930s and 1940s, many of the second generation of women home economists entered business and industry in consumer relations, marketing, product development, testing, and demonstration.[31] They acted as self-appointed mediators between consumers and producers while performing a balancing act between selling products and educating consumers now cast as women. In this position, they helped shape new consumer products and frame new meanings for these products. Within corporate and federal bureaucracies, these organization women, like their male counterparts, had little room for negotiations. However modest their contribution, these early women participated in the shaping of modern technologies.

Those who entered business or industry not only mediated but also helped to frame the meanings of new products for a female market. They presented themselves as the spokespersons for all women, whether farm wives, middle-class housewives, or urban workers. The home economists working within industry and government agencies were participants, mediators, and producers of a culture of consumption in the twentieth century. In the corporate business world, some merely acted as agents of modernity and assumed stereotypical gender roles,[32] while others explored the boundaries and tried to subvert them—albeit in modest ways.[33] At the same time, corporations realized they needed the point of view of women during their product development and began to hire women, whether or not they were trained professionals. On their part, women home economists tried to serve and mediate between two constituencies: educating companies about the needs of women consumers by presenting themselves as spokespersons on the one hand and educating women in the use of the latest household technologies on the other. Feminist research on consumer culture points to the kind of negotiations in which women's organizations have been engaged in the area between production and consumption.[34]

The case studies of the technical trajectories of bicycles, electric irons, telephones, records, and cars suggest that inventors-entrepreneurs in the early stages of technological development of consumer goods almost accidentally found new groups of users. If they failed to do so, they risked commercial failures. By contrast, the histories of utility companies show that modern entrepreneurs sought out women professionals, including home economists and women end users, in a deliberate fashion and with various degrees of success. Electricity companies were at the forefront in configuring middle-class housewives as a new market and in forming coalitions with different women's groups and professionals that could reach housewives in the intimacy of their homes. They experimented with a whole range of means, from installing test kitchens and mobilizing home economists and dietitians to investing in novel promotional campaigns in order to both reach and shape the new consumer.[35]

When gas and electric utilities were new technologies, entrepreneurs looked not just for markets but also for social groups and cultural resources in order to establish the utilities as modern and inevitable technical systems, much in the same way that Bill Gates has tried to push his computer operating system in today's world. Electricity entered the market confronted by gas, petroleum, and other fuels as well-established sources of energy. At first, the electricity companies sought to sell to factories, hospitals, and government agencies primarily for the purpose of illuminating streets and buildings at night. Electricity, unlike gas and other energy sources, is not suitable for storage, thus forcing the electricity companies to look for daytime applications, so that production costs could be spread more evenly. The electric power industry required large investments and therefore large markets to sustain itself.

At first, male marketing departments at electric companies failed to understand the working practices of the women they were trying to reach. By radically changing their marketing strategies and looking for daytime applications, the electricity companies discovered the household as their new niche in the market. Electric irons, vacuums, and stoves were developed to cater to this market. The electricity companies from California to the Netherlands discovered women. Learning from their early mistakes, the companies began to incorporate women's perspectives: they employed women to sell electrical devices and created sales environments that welcomed them as customers.

On their part, middle-class women in Britain, the Netherlands, and elsewhere established Women's Electricity Associations in an effort to mobilize the emerging technologies for women's benefit, as Mary Beard had predicted. Some women activists hoped that the new technological developments would alleviate women's household work. Others, while not against exploiting the new possibilities, were more cautious and continued to propose the redistribution of labor: they lobbied for collective laundry facilities, kitchens, and cooperatives. Nevertheless, after World War I, several middle-class women's organizations formed an alliance with the utilities as they were confronted with the new job description of the modern housewife as a full-time occupation without the help of servants.

A coalition between manufacturers and women's clubs, home economics professionals, and activists emerged in the 1920s in America and beyond, profoundly shaping the new technologies and the gender roles to go with them. During the 1890s, the National Consumers League had provided the intellectual and political framework that showed the kind of political power women as consumer activists could exercise and use to shape the direction of the old paradigm of production. From the Progressive Era onward, many women home economists made a career out of it. The alliance between these two broad interest groups—those concerned with women's issues and those concerned with corporate profit making—lasted from the 1920s until the 1960s. Then new generations of feminists began to question the whole consumption-junction framework that had employed so many women in corporate sales departments and government agencies. A new wave of the women's movement, most eloquently articulated by Betty Friedan, believed the positioning of women as consumers amounted to a modern-day nightmare of the middle-class housewife. Friedan thought to debunk the myth that women were only mothers, wives, and consumers or that working women were less happy than full-time housewives. Influenced by the work of Marx and Freud, she skillfully rephrased Vance Packard's 1957 *Hidden Persuaders* for a feminist agenda. Friedan's 1963 *Feminine Mystique* showed how the advertising industry in collaboration with women's magazines and corporate America sold home life to women as a great new market for their products by teasing out and creating women's hidden desires. However correct Friedan's analysis might have been for the 1960s, Friedan did not account for women's active participation—from the National

Consumers League to home economists—in the making of the techno-logical systems when they were still new.

CONCLUSION

Cultural practices were essential building blocks for technological developments. Women were active agents throughout the history of these practices. When the trading zone between the communities of production and consumption at the beginning of the twentieth century is opened, women are present, visible, and alive to historians. Producers and consumers were not separate actors; they shaped and constituted each other. Understanding the formation of the relationship between the two as an intricate web, we can see women in the act of creating technical artifacts—be they electric irons, urban infrastructures, or utilities.

Throughout the nineteenth and twentieth centuries, middle-class women were involved as individual end users, organized into lobby groups, or as professional spokespersons on behalf of other women. To what extent we may call them producers or coproducers of the new technologies is part of the debate that needs further exploration.[36] In some cases, women acted as mere agents of modernity, in others they initiated new technical developments, and in still others, the process was one of negotiation between more or less equal partners. When technologies were yet to find their scripts, such room for negotiation was more spacious than when the technologies began to be reinscribed as male.

For a long time, we have accepted a static economic dichotomy between production and consumption, preventing any subtle understanding of how gender formation and technological development mutually shape each other. This means that entire domains—such as cosmetics, sewing, and home economics—have not been considered appropriate fields of study for a history of technology. The static analytical dichotomies between the world of producers and that of consumers, and those between design and use, still exist. They have inhibited our ability to understand how the ideologies, structures, and identities of gender formation and technological development have shaped each other throughout the nineteenth and twentieth centuries. They prevent us from seeing how gender scripts are encoded in our technologies.[37] We could understand how gender roles are shaped through daily interactions with commonly used technologies, from color-coded lady shavers to gender-specific bike frames, which confirm

•

our gender roles whenever we use or ride them. These codes, hardened into artifacts, shape the way we understand the technologies we use as well as the gender roles we play each day.[38]

Over the last two decades or so, sociologists of science and of technology have reintroduced into our inquiries social groups as important actors in the shaping of emerging technologies. Quite separately, cultural studies focus on how representational and cultural practices encode our technologies with gendered meanings that shape design and use in profound, rather than cosmetic, ways. So far, there has been little crossover between the two. It is time to revisit Ruth Schwartz Cowan's 1982 article "The Consumption Junction," in which she questioned the dichotomous division between production and consumption prevalent in most histories of technology and argued that it would be much more fruitful to take into serious account the perspective of users rather than that of designers when examining how new technologies are adopted. So, too, we need to understand that modern manliness was revitalized through the culture of production: it took man as the maker of things and woman as the consumer of things as its guiding hope. It has become part of both history and historiography.

NOTES

1. Cynthia Cockburn, *Brothers: Male Dominance and Technical Change* (London: Pluto Press, 1983); Judy Wajcman, *Feminism Confronts Technology* (University Park: Pennsylvania State University Press, 1991); Carroll Pursell, "Toys, Technology, and Sex Roles in America, 1920–1940," in *Dynamos and Virgins Revisited: Women and Technological Change,* ed. Martha Moore Trescott (Metuchen, NJ: Scarecrow Press, 1979), 252–67; Carroll Pursell, "The Long Summer of Boy Engineering," in *Possible Dreams: Enthusiasm for Technology in America,* ed. John L. Wright (Dearborn, MI: Henry Ford Museum and Greenfield Village, 1992), 35–43; Ruth Oldenziel, "Boys and Their Toys: The Fisher Body Craftsman's Guild, 1930–1968, and the Making of a Male Technical Domain," *Technology and Culture* 38, no. 1 (1997): 60–96.

2. John M. Staudenmaier, *Technology's Storytellers: Reweaving the Human Fabric* (Cambridge, MA: MIT Press, 1985); Ruth Oldenziel, *Making Technology Masculine: Men, Women, and Modern Machines in America, 1870–1945* (Amsterdam: Amsterdam University Press, 1999); Carroll Pursell, "The Construction of Masculinity and Technology," *Polhem* 11 (1993): 206–19; Rosalyn Williams, "The Political and Feminist Dimensions of Technological Determinism," in *Does Technology Drive History? The Dilemma of Technological Determinism,* ed. Merritt Roe Smith and Leo Marx (Cambridge, MA: MIT Press, 1994), 217–35.

3. Bruce E. Seeley, "SHOT, the History of Technology, and Engineering Education," *Technology and Culture* 36, no. 4 (1995): 739–72; H. W. Lintsen and E. Homburg, "Techniekgeschiedenis in Nederland," in *Geschiedenis van de techniek in Nederland: De wording van een moderne samenleving 1800–1890,* vol. 4, ed. H. W. Lintsen et al. (Zuthpen: Walburg Pers, 1995), 255–66.

4. Joan Rothschild, ed., *Machina ex Dea: Feminist Perspectives on Technology* (New York: Pergamon Press, 1983).

5. On the cultural hierarchies of production and consumption, see Warren I. Susman, *Culture as History: The Transformation of American Society in the Twentieth Century* (New York: Pantheon Books, 1984); Warren I. Susman, " 'Personality' and the Making of Twentieth-Century Culture," in *New Directions in American Intellectual History*, ed. John Higham and Paul K. Conkin (Baltimore, MD: Johns Hopkins University Press, 1979); Leo Lowenthal, "The Triumph of Mass Idols," in *Literature, Popular Culture, and Society* (Englewood Cliffs, NJ: Prentice-Hall, 1961; reprint, Palo Alto, CA: Pacific Books, 1985); Richard Wightman Fox, "The Culture of Liberal Protestant Progressivism, 1875–1925," *Journal of Interdisciplinary History* 23, no. 3 (1993): 639–60.

6. Cynthia Cockburn and Susan Ormrod, *Gender and Technology in the Making* (London and Thousand Oaks, CA: Sage, 1993).

7. Judith A. McGaw, "No Passive Victims, No Separate Spheres: A Feminist Perspective on Technology's History," in *In Context: History and the History of Technology—Essays in Honor of Melvin Kranzberg*, ed. Stephen H. Cutcliffe and Robert C. Post (Bethlehem, PA: Lehigh University Press, 1989), 172–80.

8. For the notion of narrative productions, see Martha Banta, *Taylored Lives: Narrative Productions in the Age of Taylor, Veblen, and Ford* (Chicago: University of Chicago Press, 1993).

9. Michael Adas, *Machines as the Measure of Men: Science, Technology, and Ideologies of Western Dominance* (Ithaca, NY: Cornell University Press, 1989); Leo Marx, "The Idea of 'Technology' and Postmodern Pessimism," in *Technology, Pessimism, and Postmodernism*, ed. Yaron Ezrahi, Everett Mendelsohn, and Howard Segal, Sociology of Sciences Year Book 17 (Boston: Kluwer Academic Publishers, 1994), 11–28; Nina E. Lerman, "The Uses of Useful Knowledge: Science, Technology, and Social Boundaries in an Industrializing City," in *Women, Gender, and Science: New Directions*, ed. Sally Gregory Kohlstedt and Helen E. Longino, special issue of *Osiris* 12 (1997): 39–59; Robert Friedel, "Some Matters of Substance," in *History from Things: Essays on Material Culture*, ed. Steven Lubar and W. David Kingery (Washington, DC: Smithsonian Institution Press, 1993), 41–50.

10. Oldenziel, *Making Technology Masculine*, chap. 1; Charlotte Smith, ed., *The Woman Inventor* 1, no. 1 (1890); "Women Inventors," *Scientific American* 81 (Aug. 1899): 123; "Women as Inventors," in *Consolidated Encyclopedic Library*, vol. 10, ed. Orison Swett Marden (New York: Emerson Press, 1903), 4086–89; Otis Tufton Mason, "Woman as an Inventor and Manufacturer," *Popular Science Monthly* 47 (May–Oct. 1895): 92–103; Minnie J. Reynolds, "Women as Inventors," *Interurban Woman Suffrage Series*, no. 6 (New York: Interurban Woman Suffrage Council, 1908) (reprinted in *New York Sun*, Oct. 25, 1908); James Johnson, "Women Inventors and Discoverers," *Cassier's Magazine: An Engineering Magazine* 36, no. 6 (Oct. 1909): 548–52; Mary Ritter Beard, "Inventions Are for Men," letter to the editor, *New York Times*, Aug. 17, 1915, p. 8, col. 8. For intriguing episodes between individual women inventors and women activists, see Anne L. Macdonald, *Feminine Ingenuity: Women and Invention in America* (New York: Ballantine Books, 1992). Autumn Stanley, *Mothers and Daughters of Invention: Notes for a Revised History of Technology* (Metuchen, NJ: Scarecrow Press, 1993), provides the most comprehensive listing of the thousands of women inventors in the United States.

11. Oldenziel, *Making Technology Masculine*; Macdonald, *Feminine Ingenuity*.

12. Virginia Grant Darney, "Women and World's Fairs: American International Expositions, 1876–1904" (Ph.D. diss., Emory University, 1982); Jeanne Madeline Weimann, *The Fair Women: The Story of the Woman's Building, World's Columbian Exposition, Chicago, 1893* (Chicago: Academy Chicago, 1981). See also Maria Grever and Berteke Waaldijk, *Feministische openbaarheid:*

De nationale tentoonstelling van vrouwenarbeid in 1898 (Amsterdam: IISG/IIAV, 1998).

13. Beard, "Inventions Are for Men"; Mary Ritter Beard, *Woman's Work in Municipalities* (1915; reprint, New York: Arno Press, 1972); Mary Ritter Beard, *Woman as Force in History: A Study in Traditions and Realities* (New York: Macmillan, 1946).

14. As cited in Steven Lubar, "Men/Women/Production/Consumption," in *His and Hers: Gender, Consumption, Technology,* ed. Roger Horowitz and Arwen Mohun (Charlottesville: University of Virginia Press, 1998), 7–37. See also Nina E. Lerman, "From 'Useful Knowledge' to 'Habits of Industry': Gender, Race, and Class in Nineteenth Century Technical Education" (Ph.D. diss., University of Pennsylvania, 1993).

15. Ruth Schwartz Cowan, "The Consumption Junction: A Proposal for Research Strategies in the Sociology of Technology," in *The Social Construction of Technological Systems: New Directions in the Sociology and History of Technology,* ed. Wiebe E. Bijker, Thomas P. Hughes, and Trevor J. Pinch (Cambridge, MA: MIT Press, 1987), 261–80.

16. Joy Parr, "Shopping for a Good Stove: A Parable about Gender, Design, and the Market," in Horowitz and Mohun, *His and Hers,* 165–88; Joy Parr, "What Makes Washday Less Blue? Gender, Nation, and Technology," *Technology and Culture* 38, no. 1 (1997): 153–86.

17. Ronald Kline and Trevor Pinch, "Users as Agents of Technological Change: The Social Construction of the Automobile in the Rural United States," *Technology and Culture* 37, no. 4 (1996): 763–95; Reynold M. Wik, "The Early Automobile and the American Farmer," in *The Automobile and American Culture,* ed. David L. Lewis and Laurence Goldstein (Ann Arbor: University of Michigan Press, 1983), 37–47.

18. Bryan Pfaffenberger, "Fetishised Objects and Humanised Nature: Toward an Anthropology of Technology," *Man* 23 (June 1988): 236–52. See also on this point Langdon Winner, "Do Artifacts Have Politics?" *Daedalus* 109, no. 1 (1980): 121–31; and Madeleine Akrich, "The De-Scription of Technical Objects," in *Shaping Technology/Building Society: Studies in Sociotechnical Change,* ed. Wiebe E. Bijker and John Law (Cambridge, MA: MIT Press, 1992), 205–24.

19. Akrich, "The De-Scription of Technical Objects"; Bryan Pfaffenberger, "Technological Dramas," *Science, Technology, and Human Values* 17, no. 2 (1992): 282–312; and Bryan Pfaffenberger, "User Representations: Practices, Methods, and Sociology," in *Managing Technology in Society: The Approach of Comparative Technology Assessment,* ed. Arie Rip, Thomas Misa, and Johan Schot (London: Pinter, 1995), 167–84.

20. W. Bernard Carlson, "Artifacts and Frames of Meaning: Thomas A. Edison, His Managers, and the Cultural Construction of Motion Pictures," in Bijker and Law, *Shaping Technology/Building Society,* 175–200, quotation on 176.

21. On bicycles, see Wiebe Bijker and Trevor Pinch, "The Social Construction of Artifacts," in Bijker, Hughes, and Pinch, *The Social Construction of Technological Systems,* 17–50. I have also benefited from Dirk Stemerding's interpretation of Wiebe E. Bijker, *Of Bicycles, Bakelites, and Bulbs: Toward a Theory of Sociotechnical Change* (Cambridge, MA: MIT Press, 1995), 19–100: "Een sociologische kijk op technologie," in *Technologie en samenleving,* ed. H. Achterhuis et al. (Heerlen: Open Universiteit, 1995), 47–72. See also Ellen Gruber Garvey, "Reframing the Bicycle: Advertising-Supported Magazines and Scorching Women," *American Quarterly* 47, no. 1 (1995): 66–101.

22. Louis Carlat, " 'A Cleanser for the Mind': Marketing Radio Receivers for the American Home, 1922–1932," in Horowitz and Mohun, *His and Hers,* 115–38; Susan Smulyan, *Selling Radio: The Commercialization of American Broadcasting, 1920–1934* (Washington, DC: Smithsonian Institution Press, 1994); Su-

san J. Douglas, *Inventing American Broadcasting, 1899–1922* (Baltimore, MD: Johns Hopkins University Press, 1987).

23. On cars, see Kline and Pinch, "Users as Agents of Technological Change"; Virginia Scharff, *Taking the Wheel: Women and the Coming of the Motor Age* (Albuquerque, NM: University of New Mexico Press, 1992), chap. 4; Rudi Volti, "Why Internal Combustion?" *American Heritage of Invention and Technology* 6 (fall 1990): 42–47; Gijs Mom, *De geschiedenis van de auto van morgen* (Deventer: Kluwer Bedrijfsinformatie, 1997).

24. Claude S. Fischer, " 'Touch Someone': The Telephone Industry Discovers Sociability," *Technology and Culture* 29, no. 1 (1988): 32–61; Claude S. Fischer, *America Calling: A Social History of the Telephone to 1940* (Berkeley and Los Angeles: University of California Press, 1992). For contemporary cases, see Lana F. Rakow and Vija Navarro, "Remote Mothering and the Parallel Shift: Women Meet the Cellular Telephone," *Critical Studies in Mass Communication* 10, no. 2 (1993): 144–57; Lana F. Rakow, *Gender on the Line: Women, the Telephone, and Community Life* (Urbana: University of Illinois Press, 1992); Lana F. Rakow, "Women and the Telephone: The Gendering of Communications Technology," in *Technology and Women's Voices: Keeping in Touch*, ed. Cheris Kramarae (New York: Routledge, 1988); Ann Moyal, "The Gendered Use of the Telephone: An Australian Case Study," *Media, Culture, and Society* 14, no. 1 (1992): 51–72. For the point of view of women workers in the telephone industry, see Kenneth Lipartito, "When Women Were Switches: Technology, Work, and Gender in the Telephone Industry, 1890–1920," *American Historical Review* 99, no. 4 (1994): 1074–111; Michele Martin, *"Hello, Central?" Gender, Technology, and Culture in the Formation of Telephone Systems* (Montreal: McGill-Queen's University Press, 1991); Angela E. Davis, " 'Valiant Servants': Women and Technology on the Canadian Prairies, 1910–1940," *Manitoba History* 25 (1993): 33–42; Venus Green, "The Impact of Technology upon Women's Work in the Telephone Industry, 1880–1890" (Ph.D. diss., Columbia University, 1990).

25. Kathryn Kish Sklar, "The Consumers' League, 1898–1918," in *Getting and Spending: European and American Consumer Societies in the Twentieth Century*, ed. Susan Strasser, Charles McGovern, and Matthias Judt (Cambridge: Cambridge University Press, 1998), 17–36; Jackie Dirks, "Righteous Goods: Women's Production, Reform Publicity, and the National Consumers' League, 1891–1919" (Ph.D. diss., Yale University, 1996).

26. Karen J. Blair, *The Clubwoman as Feminist: True Womanhood Redefined, 1868–1914* (New York: Holmes and Meier, 1980); Steven M. Buechler, *The Transformation of the Woman Suffrage Movement: The Case of Illinois, 1850–1920* (New Brunswick, NJ: Rutgers University Press, 1986); Mina Carson, *Settlement Folk: Social Thought and the American Settlement Movement, 1885–1930* (Chicago: University of Chicago Press, 1990); Judith Ann Trolander, *Professionalism and Social Change: From the Settlement House Movement to Neighborhood Centers, 1886 to the Present* (New York: Columbia University Press, 1987).

27. Beard, *Woman's Work in Municipalities*.

28. Sarah Stage, "Ellen Richards and the Social Significance of the Home Economics Movement," in *Rethinking Home Economics: Women and the History of a Profession*, ed. Sarah Stage and Virginia B. Vincenti (Ithaca, NY: Cornell University Press, 1997), 17–18.

29. For a provocative inquiry into the area of crossover between public and private technical cultures, see this volume, chap. 8. See also Oldenziel, *Making Technology Masculine*, chap. 5. For an international comparison of the relationship between women engineers and the women's movement, see Annie Canel, Ruth Oldenziel, and Karin Zachmann, eds., *Crossing Boundaries, Building Bridges: Comparing the History of Women Engineers, 1870s–1990s* (London: Harwood Academic Publishers, 2000).

30. Nancy F. Cott, *The Grounding of Modern Feminism* (New Haven, CT: Yale University Press, 1987).

31. The rest of this section is based on the essays in Stage and Vincenti, *Rethinking Home Economics*.

32. Ronald R. Kline, "Agents of Modernity: Home Economists, Technology Transfer, and Rural Electrification in the United States, 1925–1950," in Stage and Vincenti, *Rethinking Home Economics*, 237–52.

33. Carolyn Goldstein, "Mediating Consumption: Home Economics and American Consumers, 1900–1940" (Ph.D. diss., University of Delaware, 1994); Carolyn Goldstein, "Part of the Package: Home Economists in the Consumer Product Industries, 1920–1940," in Stage and Vincenti, *Rethinking Home Economics*, 271–97; Lisa Mae Robinson, "Safeguarded by Your Refrigerator: Mary Engle Pennington's Struggle with the National Association of Ice Industries," in Stage and Vincenti, *Rethinking Home Economics*, 253–70.

34. To name only two of the many new works: Jackie Dirks, "Righteous Goods"; Dana Frank, *Purchasing Power: Consumer Organizing, Gender, and the Seattle Labor Movement, 1919–1929* (Cambridge: Cambridge University Press, 1994). For a useful overview, see Victoria De Grazia and Ellen Furlough, eds., *Sex of Things: Gender and Consumption in a Historical Perspective* (Berkeley and Los Angeles: University of California Press, 1996). While many studies focus on women, they usually approach the sources from a business point of view; nevertheless, they are important: Susan Strasser, *Satisfaction Guaranteed: The Making of the American Mass Market* (New York: Pantheon Books, 1989); Jennifer Scanlon, *Inarticulate Longings: The Ladies' Home Journal, Gender, and the Promises of Consumer Culture* (New York: Routledge, 1995).

35. Gregory Field, " 'Electricity for All': The Electric Home and Farm Authority and the Politics of Mass Consumption," *Business History Review* 64 (spring 1990): 32–60; Carroll Pursell, "Domesticating Modernity: The Electrical Association for Women, 1924–1986," *British Journal for the History of Science* 112, no. 32 (Mar. 1999): 47–67; Peter van Overbeeke, "Koken op gas of op electriciteit: Een strijd om de vrouw," in *Schoon genoeg*, ed. Ruth Oldenziel and Carolien Bouw (Nijmegen: SUN, 1998), 127–58.

36. See Roger Horowitz and Arwen Mohun, introduction to *His and Hers*, 1–6; I thank Judy Wajcman for engaging me in discussion on this issue (conversation with author, Jan. 17, 1999).

37. For more on the gendering of birth control, see this volume, chap. 11.

38. Ellen Oost, "Over 'vrouwelijke' en 'mannelijke' dingen," in *Vrouwenstudies in de jaren negentig: Een kennismaking vanuit verschillende disciplines*, ed. Margo Brouns, Mieke Verloo, and Marianne Grünell (Bussum: D. Continho, 1995), 289–313; Nelly Oudshoorn, *Genderscripts in technologie: Noodlot of uitdaging?* (Enschede: Universiteit Twente, 1996).

EIGHT

What Difference Has Feminism Made to Engineering in the Twentieth Century?

PAMELA E. MACK

Until the last quarter of the twentieth century women were very scarce in engineering, so the impact of feminism on engineering might seem like a topic with a very short history. However, women engineers did make contributions to the profession in the late nineteenth and early twentieth centuries. Furthermore, somewhat broader definitions of feminism and of engineering bring to light very significant influences of the women's reform movement of the first half of the twentieth century on industrial and municipal engineering. The second feminist movement brought a new emphasis on equal rights and also a new kind of difference feminism, which focused its challenge to engineering on the level of assumptions rather than on the level of practices. Second-wave feminism's emphasis on equal rights led to more opportunities for women in engineering but did not significantly alter the nature of the profession. Specifically, the new feminist critique of engineering had little effect on practitioners in the field, who accepted at most equal-rights feminism. Neither did women engineers find in feminism a source of new ideas and special job opportunities the way they had in the reform movement decades earlier. This chapter follows these changing historical interactions between feminism and engineering, both in theory and in practice. Here and elsewhere, I argue that women and women's movements have had an impact on engineering throughout the twentieth century, but that the character of these contributions changed dramatically with the coming of second-wave feminism in the 1960s.[1]

FEMINIST THEORIES

Feminism broadly defined has changed over time and can be categorized in many ways. For the purposes of this chapter it is enough to
distinguish between difference feminism and equal-rights feminism.
The critique of engineering developed by late-twentieth-century feminists differed in important ways from earlier feminist ideas.

Equal-rights feminists believe that the differences between men and
women are not significant and that the goal should be equal treatment
of men and women. Early-twentieth-century equal-rights feminists
mostly made philosophical arguments for rights for women. But some
women who moved into formerly male professions in this early period
took a quiet equal-rights stance, hoping that if they asked for no special consideration because they were women they could prove themselves worthy of equal treatment.[2]

Another approach, however, had more impact in first-wave feminism. Difference feminists believe that women have something special
to contribute because they are different from men, either essentially
or because of their upbringing and/or cultural history.[3] The experiences of middle-class American women in the late nineteenth century
led some women leaders to develop a vision of what they could contribute as women, and that particular vision, sometimes called social
feminism, provided themes for the women's reform movement through
the first half of the twentieth century. By the 1890s women had made
new opportunities for themselves in the public sphere, based in large
part on the argument that women were essentially different from
men—more moral, more concerned with protecting the weak, having
the special skills of housekeeping. Women in the Progressive movement argued that those differences allowed them to make unique contributions especially in the area of undoing the harm wreaked by un-
controlled capitalism and technological change. These Progressive
women made an impact on engineering by working toward regulation,
particularly of the workplace and of factories producing hazardous
wastes. Some also gained considerable technical expertise in fields
varying from economics to garbage disposal.[4]

The influence of traditional difference feminism faded considerably
in the middle of the twentieth century. When a second wave of feminism developed in the late 1960s and early 1970s, its political wing
made equal rights the central policy goal and explicitly rejected many
of the old arguments for difference. Many second-wave feminists argued that women were not necessarily different from men except in a

few narrow areas (such as childbirth and breast-feeding) and that women should fight for a society where equal treatment of men and women would result in equal participation in all aspects of society.

While second-wave feminism centered its political action on equal rights, difference feminism quickly became important in feminist theory. These approaches had some roots in the earlier reform arguments, though the new feminist theorists rejected many of the old approaches. Some women, like Rachel Carson, bridged the period between the reform tradition and a new, more radical critique of a world based on money, power, and technology.[5] In making that new critique, however, second-wave difference feminists wanted to go much further than the reform tradition. They criticized not just the organization of production but the relationship between human beings and nature.

Early second-wave feminist theory did not make technology a primary issue, but over time some feminist theorists came to see technology as central to the workings of patriarchy. Carolyn Merchant's 1980 book *The Death of Nature* perhaps made the strongest early argument, though she drew her material more from the history of science than from the history of technology. Merchant examined "the formation of a world-view and a science that, by reconceptualizing reality as a machine rather than as a living organism, sanctioned the domination of both women and nature."[6] This, together with Lynn White's argument that Christianity led to the exploitation of nature, gave rise to theories that saw science and technology as closely interwoven with the subjugation of women.[7] According to this approach women represented an alternative set of values to a patriarchal system based on the exploitation of nature, women, and other groups given second-class status. However, this approach made many second-wave feminists uncomfortable because it separated women from the basis of power in society as it was and is.

These ideas saw more development by ecofeminists, particularly Christian ecofeminists, than by scholars in the social studies of technology.[8] The conservation movement had been in many ways conservative, whereas the new environmental movement critiqued industrial capitalism much more directly. The feminist and environmental movements of the 1970s borrowed from each other, though their theoretical interconnections proved controversial. Some theorists argued that women must take the lead in abandoning patriarchal values and rediscovering women's special relationship with nature.[9] Others asserted that any argument based on women's difference would consign women to inequality and argued instead that the hierarchical dualism between nature and culture should be questioned in the same ways that feminists had questioned the dualism between men and women.[10] On the whole, ecofeminists and feminists in the environmental movement tended to image technology, particularly big, science-based technology, as male exploitation.

A few feminist sociologists pursued the idea that engineering has a traditional association with masculine values into a more specific critique of engineering. Sally Hacker argued that mind/body dualism was particularly strong in the culture of engineering, and therefore technical skills were constructed as the opposite of skills of nurturance and responsiveness.[11] Hacker leaned in the direction of the idea that engineering would be improved by an injection of women's values, whereas another theorist, Judy Wajcman, leaned toward the perspective that the masculine character of engineering was simply a strategy to keep women out of power. Wajcman pointed out that the definition of what is masculine changes to meet the needs of men, for example, sometimes privileging physical force and sometimes intellectual rationality. She concluded: "No matter how masculinity is defined according to this ever-adaptable ideology, it always constructs women as ill-suited to technological pursuits."[12] A study conducted in 1986 and published in 1991 by Gregg Robinson and Judith McIlwee looked for this in practice, focusing on the idea that the culture of engineering privileges masculinity. They concluded that women engineers had more success in workplaces where engineers had less power because in situations dominated by engineers, workplace culture "equates professional competence with 'masculinity.'"[13] A British study similarly argued: "What is especially difficult about engineering, compared to any other profession, is the cult of 'physical masculinity' connected with the work. This physical masculinity is also heavily sexualized, often overtly."[14]

These critiques had little impact inside engineering because feminist practitioners did not gain recognition and influence. Searching for the word "feminism" in the engineering literature turns up two pockets of feminist theory in fields somewhat on the edge of engineering. In computer science feminists pointed out how education is gendered.[15] In the area of reproductive technology in particular and medical ethics more generally, an extensive feminist literature also developed.[16] In both cases scholars from other fields were usually the first to raise feminist arguments, not engineers and other practitioners, but feminist analysis had significant impacts in each of these fields. However, this process has not taken place in more mainstream engineering.

In mainstream engineering, discussion of feminism was watered down into a few positive statements about the possible contributions of women. Samuel Florman established himself in public culture as a spokesperson for engineering with the 1976 publication of his book *The Existential Pleasures of Engineering.*[17] He wrote an article in *Harper's* in 1978 in which he argued that "talented young women were avoiding engineering because they perceived other professions as a more direct route to political power. . . . if you are smart enough to be an engineer, you're too smart to be an engineer."[18] By the 1986 publication of *The Civilized Engineer,* Florman had switched strategies and was writing about the "fresh and valuable elements" women could bring to the profession, particularly a more humanistic approach.[19] One of Florman's major theses is that humanistically educated engineers are more creative; but women engineering students not surprisingly felt nervous about his argument for difference.[20] Sometimes when engineers give speeches about the future of engineering, they mention in passing the argument that women tend to have skills such as working well with others and integrating diverse information that fit new trends in engineering work,[21] but there is little evidence that this rhetoric has led to any advantage for women in hiring and promotion.[22]

Despite the limited interest of practitioners in the feminist critique of the gendered culture of engineering, some sociologists have tried to determine whether women do engineering differently or whether engineering might be different if it were less identified with traditionally masculine values. The evidence for women bringing new ideas into engineering is mixed, at best. A 1986 study interviewed women engineers about "engineers' attitudes to technological development and change, to see how far women who have been socialized into a masculine profession adopt masculine value systems and how far they retain female value systems, bringing a different perspective to traditionally masculine work." The authors found only a few signs that women engineers had different values. In particular, they saw very little questioning of definitions of progress, though they found some notable idealism among the women engineers they interviewed. They concluded that when women reached a critical mass in the workplace they would be more likely to raise gender issues.[23]

In a study published in 1992, Knut Sørensen sought a broader empirical answer to the question of whether women do technology differently. Although his subjects were Norwegian women and men and his workplace sample included both scientists and engineers working in research and development (R&D), his analysis is valuable for its careful evaluation of feminist arguments about difference. Sørensen carefully avoids claiming that women are essentially different, but he argues that women have different cultural resources that they may bring to a field: "In their perception of important challenges to R&D, women are more inclined to include reproductive considerations; they have a caring, other-oriented relationship to nature and to people, an integrated, more holistic and less hierarchical world-view; a less competitive way of relating to colleagues and a greater affinity to users; or generally—to use Hilary Rose's elegant metaphor—an ability to bring together the knowledges of hand, brain and heart. The resulting ideal type is to be juxtaposed to the similarly stereotyped 'masculine counter-values': domination of nature, objectivation of research, competitiveness, acceptance of authority, and the like." Sørensen's results suggest that women engineering students are more likely than men to claim "caring values." However, these differences did not show up in the work norms of his sample of young researchers, suggesting that socialization reduces differences by the time women finish their schooling. Nonetheless, Sørensen suggests that women in R&D find opportunities to use their values in choosing their field and their research problems, pointing to the differences in the percentages of women in different fields of engineering.[24] There is some evidence that the same patterns hold in the United States; my own survey of students in a class for freshman engineers found women students much more likely than men students to believe that social issues are important in engineering.[25] However, these studies examine only stated values, not behavior. A 1996 study of women managers showed that even when both women and men paid lip service to ideas of women's management style, successful women tended to follow male styles.[26]

Feminist theory suggests the possibility that women might bring new ideas and approaches to engineering, as they did in the earlier women's reform movement, but there is little evidence that this possibility has been realized. It is not simply that the number of women engineers must reach a critical mass before they will have a significant impact on engineering, because the women's reform movement had considerable impact on engineering at a time when the proportion of women engineers was even smaller. Perhaps the problem has been that feminist critiques of technology rejected such fundamental assumptions of the field that it seemed impossible to incorporate those critiques into engineering practice. I suspect, however, that the new ideas of the reformers in the early twentieth century seemed almost as radical at the time. Just as the earlier women's reform movement led women like Lillian Gilbreth to innovative and influential new approaches in their technical work, in many fields feminism has led women to ask new questions that scholars in that field have found compelling. However, that has happened very little in engineering. Occasional articles in liberal journals, such as one entitled "Want Your Machine to Work? Get a Woman to Design It," have suggested the possibility of similar patterns in engineering, but little has come of it.[27]

WOMEN ENGINEERS

The vast majority of the literature on women and engineering deals with the small number of women in engineering and with how more women might be encouraged to study engineering. This enterprise involves equal-rights feminism, since it involves women working for equality, though many participants are uncomfortable with the word "feminism." Promoters of women in engineering have steered away from the broader feminist critique of engineering based on difference feminism.

Even in the early twentieth century a small number of women went to engineering school and pursued careers in engineering despite discrimination and the increasingly masculine culture of professional engineering.[28] The careers of pre–World War II women engineers fall into two patterns. Some did relatively conventional work, hoping to win acceptance by a strategy of overqualification, stoicism, and modesty.[29] Opportunities for these women often came through family connections, usually by working in a family business.[30] Others were motivated by reform impulses and moved back and forth between conventional engineering work and women-oriented or reform-oriented work. Reform activities seem to have provided women like Lillian Gilbreth with motivation and support for technical work, with job opportunities, and with a source of new approaches that often led to creative careers.[31]

As the women's reform movement gradually lost visibility through the middle years of the century, women continued to pursue engineering. The primary motive for new opportunities for women in engineering during World War II and for the continuation of some encouragement for women after the war was not feminism in any form but worry about a shortage of "manpower" for wartime and Cold War projects.[32] New job opportunities for women in World War II gave more women a start, but women were almost always limited to low-level jobs, often as engineering aides.[33] Women studied engineering in somewhat larger numbers after the war than in the decades before the war, though their percentage did not go up because the number of men was growing just as rapidly. The 1950 census showed 6,475 female and 518,781 male engineers (1.2 percent).[34] Arminta Harness, who studied engineering immediately after the war, said that women flooded the engineering schools, but "[w]hen I graduated, industry informed me that they were not interested in hiring a woman engineer." She made a career as an engineer in the Air Force instead.[35] Many women engineers were isolated and did not want to be seen as different from their male colleagues, but some did attempt to argue for a place for women in engineering. Informal gatherings of women engineers began among "[w]omen who had made sizeable and significant contributions to the war effort [and] were not willing to retreat to old norms and mores."[36]

Despite the slow change in the proportion of women, attitudes were changing in the 1950s and 1960s, as Margaret Rossiter has shown. In 1954 the Women's Bureau of the U.S. Department of Labor published a bulletin entitled *Employment Opportunities for Women in Professional Engineering*. Unlike the situation in most fields of science, women engineers did not find opportunities to be better in government and nonprofit institutions than in industry in the 1950s; women made up 0.5 percent of the engineers in industry in 1956–58.[37] Rossiter argues that women's associations were only a partial palliative, but given the lack of visibility of women engineers, the founding of the Society of Women Engineers in 1950 represented a significant step forward. By 1958 the society had 510 members, which suggests at least that women felt a need for such a group.[38]

These efforts led to a continuing increase in the number of women engineers, but since the number of men continued to rise and women were often excluded or limited by quotas, there was little change in the proportion of women.[39] The Civil Rights Act of 1964 outlawed sex discrimination in many areas (though not education), but without a political movement sponsoring challenges under the new law it had little initial effect.[40] A 1965 analysis of the lack of women engineers focused on the problems of overt discrimination.[41] Through the 1960s and even into the 1970s many of the elite engineering schools struggled over the issue of coeducation; the Massachusetts Institute of Technology built its first dorm for women in 1960, the Georgia Institute of Technology admitted women to all programs in 1965, and the California Institute of Technology decided to admit women as undergraduates in 1968. Even in the early 1970s Rensselaer Polytechnic Institute had an informal policy of admitting no more than 140 women a year.[42]

The percentage of women in engineering started to increase noticeably in the 1970s, and the increase became significant in the 1980s, in part because of changes within engineering. In the period from 1968 to 1972 women scientists and engineers began to talk about fairness and the need for change.[43] A major slump in engineering jobs occurred in the early 1970s because of the termination of Project Apollo and cuts in military R&D. Nance Dicciani, with a master of science degree in chemical engineering from the University of Virginia, said of her experiences on the job market in 1970: "Women engineers were not even wanted, let alone sought after. I can remember interviews where people actually said, 'the only women we hire here are secretaries and we don't have any openings so you needn't fill out an application.'"[44] However, the situation changed quickly. The number of young white men choosing to study engineering dropped in response to the poor job market, and engineering schools recruited women and minorities to keep up student numbers and quality.[45] Soon, federally mandated affirmative action began to make a difference in industries that relied heavily on government contracts; in 1974 electrical engineer (and Olympic swimmer) Lynn Colella Bell found that "many of the larger companies specifically wanted to interview women and there were only a few of us to go around."[46] Organizations also began to consider the issue of women engineers: the American Society of Engineering Education formed a Task Force on Women in 1974 (later renamed the Committee on Women in Engineering).[47]

The broader feminist movement probably had at least as substantial an influence as action by women engineers. Feminists fought for laws and regulations against patterns of discrimination that had limited or cut off many women's careers. They filed lawsuits against discrimination under various laws and executive orders and publicized both positive results and the slowness of change.[48] They fought for tougher laws and won passage of the Equal Employment Opportunity Act of 1972 and Title IX that same year. They worked for affirmative action programs as a method of remedying a past pattern of discrimination, and from the mid-1970s to the mid-1980s such programs made a significant difference for women in universities and in industries heavily dependent on government contracts.[49] Feminism also changed the broader culture. Many young women students in the 1970s were emboldened to argue that women could do anything, and with the help of affirmative action they were able to chip away at the barriers to women's participation in engineering.

By the late 1970s and early 1980s women had made progress (fig. 8.1). The percentage of engineering students who were women was rising significantly for the first time. At least some male engineering professors became concerned about improving opportunities for women, particularly men who saw their own daughters struggling to break into the professions.[50] Women still faced a hostile classroom climate in some of their engineering classes, but by the early 1980s affirmative action and

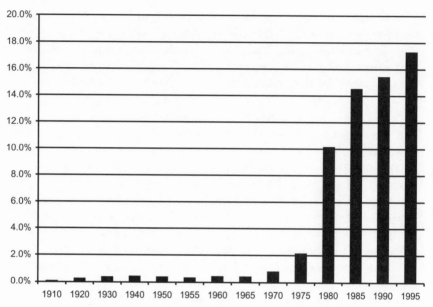

Figure 8.1 Bachelor's degrees in engineering received by women

the increasing number of women had broken through many of the more obvious barriers (such as the lack of "ladies rooms").[51] Feminism had led to more jobs for women in engineering not by creating special opportunities for women but by winning legal and administrative changes that required more equal treatment of women.

The attitude of women engineers was changing as well. Previously, many women engineers had chosen the strategy of trying to appear "genderless," seeking to be seen only as professionals.[52] Engineers are not usually the kind of people who choose to fight the system, perhaps because they mostly work within a fixed system of procedures and "laws of science" and so they tend to imagine the social world as equally fixed. As feminism shifted the values of the society around them, however, women engineers began to change their expectations. When young women with feminist ideas first began to move into engineering in the 1970s, they found that some of the women already in the field had the attitude that they had made it the hard way and other women could too.[53] However, even many of these senior women gradually shifted their attitudes. One woman with a long career in engineering graphics, Margaret Eller, exemplified the mixing of old and new attitudes in a 1990 interview. She said that of course she was propositioned at work, but "I just ignored it, which I think a lot of women could do today." Yet she also said that if women "had been more aware of each other—networked—advances in combating sexual harassment could have occurred sooner."[54] When they joined together, women engineers still sought more to encourage each other than to change the profession, but their expectations of how they should be treated did change.

It quickly became clear that providing equal opportunity as mandated by law was not by itself enough to bring women into engineering in large numbers. By the early 1980s, analysts tended to assume that discrimination was no longer a significant problem, though a 1987 study showed that 27 percent of the male engineers surveyed agreed with the statement "The ability of women engineers is less compared to men."[55] In any case, subtle prejudice was still common in engineering, sometimes much to the surprise of young women who did not expect awkwardness just because they were women.[56] In school subtle prejudice took such forms as giving less attention to women students, assuming particular knowledge, and using examples from traditional teenage boy culture, such as presupposing that students knew the details of how an automobile engine worked.[57] In the workplace subtle prejudice could take such forms as doing business at informal gatherings where women were not welcome, sexist jokes and pornographic pictures in work areas, and outright resistance by blue-collar workers to taking orders from women.[58]

Rather than tackle subtle prejudice head on, studies of how to encourage more women to study engineering came to emphasize networking, mentoring, and career development programs. Organizations such as the Women in Engineering: Program Advocates Network, the Society for Women Engineers, and the Women in Engineering Program of the Institute of Electrical and Electronic Engineers provided information to encourage young girls to consider careers in engineering, mentors for women undergraduates, and networking for working women.[59] These programs aimed mostly at making women feel more comfortable in engineering, or at least giving them someone to turn to for advice and comfort when they felt uncomfortable.

The feminist critique of the masculine style of engineering received occasional attention within this effort. Some studies showed that older firms and older industries tended to have stronger old-boy networks and climates in which women felt more isolated.[60] However, Robinson and McIlwee's study of the culture of engineering described what they call a "technical locker room" style in young high-tech companies, in which women felt very uncomfortable. In this culture, "aggressive displays of technical competence [were] the criteria for success."[61] They contrasted this with more bureaucratic firms in which women fared better because of affirmative action, because engineers had less decision-making power in those firms and because the corporate culture outweighed the engineering culture. Robinson and McIlwee concluded: "Men are not better engineers, but they are better at *appearing* to be better engineers in a *male-defined* way." [62]

Responses by engineers to Robinson and McIlwee's study tended to restate the message in ways that put the responsibility on women. An article by an engineering graduate student that borrowed heavily from Robinson and McIlwee stated: "In workplaces where engineers hold power, a culture of engineering based on a male style of interaction prevails. Women fail to understand the unwritten rules of conduct and, as a result, fail to effectively promote their strengths." [63] Or, in the words of an article written by women and published in *Science:* if women are isolated they "may be slow to learn the 'unwritten rules' that lead to corporate success." [64] The Society of Women Engineers even published an article in 1993 entitled "Learning to Compete with Women." [65] The alternative view, that "there is little reason to believe that existing cultural norms are necessary to the pursuit of excellence in science and engineering," has received little attention. [66]

While progress has been slow for women in engineering overall, there is considerable variation among subfields. Chemical engineering was traditionally the field with the most women engineering majors, though civil engineering was more popular for a period in the early to mid-1970s. [67] The fields of engineering that have the highest proportion of women today are bioengineering (mostly developing medical devices) and environmental engineering, both highly interdisciplinary fields in which the masculine culture of engineering may be somewhat diluted. In my own institution, in 1999 women made up more than 50 percent of the incoming graduate students in environmental engineering. [68] Table 8.1 shows the percentages of bachelor's degrees going to women in some major

TABLE 8.1 *Percentage of Women in Major Fields of Engineering*

FIELD	WOMEN (%) 1983	WOMEN (%) 1993	CHANGE IN FIELD SIZE (%)	CHANGE IN NUMBER OF WOMEN (%)
Aeronautical/astronautical engineering	8	11	29	84
Chemical engineering	21	32	−43	−13
Civil engineering	13	22	−9	20
Electrical engineering	14	18	2	18
Industrial engineering	26	29	−6	4
Mechanical engineering	9	11	−8	10
Materials/metal engineering	21	21	−13	−10
Other engineering	14	18	−38	−25

fields of engineering in 1983 and 1993.[69] The column "change in field size" shows the change in the total number of degrees granted in that field, and the column "change in number of women" shows the percentage change in the total number of women receiving degrees in that field.

For example, the percentage of degrees in chemical engineering going to women increased, but the total number of degrees in chemical engineering declined sharply, and even with an increase in percentage, the number of women receiving degrees in that field declined. A finer division is required to draw many conclusions, but chemical engineering probably includes most students aiming at careers in environmental engineering (industrial and management engineering is the field pioneered by Lillian Gilbreth). One analysis of such figures argues that young women often do not want to be seen as masculine, and they perceive fields like chemical engineering to be less masculine than, for example, mechanical engineering.[70] However, the difference could also be explained by the argument that young women are more likely than young men to be interested in fields that they see as more involved with people and as more idealistic.

Women have found an increasing number of job opportunities in engineering, according to a pattern quite different from the early twentieth century. In the earlier period women used the women's reform network to develop some job opportunities particularly open to women, such as industrial health. Second-wave feminism fought instead for women to compete equally for the same jobs as men. This approach has been successful to an extent that would have been beyond the wildest dreams of women with technical expertise in the early twentieth century. However, because this approach was based on an argument from equality, it has made any discussion of how women might be different seem very risky.[71] It has therefore been very difficult for women to use the feminist critique of engineering to reform engineering. In the end, if the difference feminism makes is defined by whether increasing participation by women has changed the engineering profession, it is simply too early to answer the question.

CONCLUSIONS

When women worked for a larger role in public life, it made a difference whether they did so by arguing that they had a special role or by arguing for equal treatment. My interest in the tension between difference and equal-rights feminism is not a philosophical one but a practical one: what results came of women taking these two different approaches for the field of engineering? Both approaches to feminism coexisted throughout the twentieth century, though difference feminism was the more common strategy of women seeking to have a political influence in the first half of the twentieth century, whereas equal-rights feminism played that role for second-wave feminism. Both kinds of feminism had advantages and disadvantages for women seeking to influence and participate in the public and professional worlds. The two kinds of feminism provided different opportunities both for individual women and for society as a whole.

Difference feminism provided a way to get a foot in the door, by arguing that women's participation was needed on particular topics. It could also provide women with a chance to make innovative contributions to whatever field they were entering, since by definition they brought new ideas and new approaches. However, difference feminism tended to limit opportunities for women to topics or fields that were perceived as particularly suitable for women. The theory of complementary or separate but equal roles for men and women almost inevitably resulted in unequal treatment for women.

Equal-rights feminism insisted that women gain access to the mainstream. Women had the opportunity to work in any area that they chose, not just those that fit some theory of women's special role. However, equal-rights feminism made it difficult to challenge existing male-defined ways of doing things because women accepted on the basis of equality did not want to call attention to themselves as different from men. Even a better idea was dangerous if it might be seen as particularly female and therefore point to women as different and potentially unequal.

Feminism has the most impact on a field when it creates opportunities for women, either through jobs or by providing a source for innovative ideas. Early-twentieth-century difference feminism clearly opened up opportunities for middle-class women to deal with issues relating to factories and municipal engineering and to make a difference as reformers, though few women made it into the mainstream of engineering. In the early twentieth century, women did not form an interest group separated from engineering and technical expertise. Women reformers used technical expertise and alliances with experts, and women experts found career opportunities and made creative contributions to their fields because of their reform beliefs. Women engineers could try to be accepted as equal in the culture of engineering, and some did attempt that, but women found themselves more welcome in the reform enterprises. Precisely because they did something different, some women with reform interests made particularly significant contributions to technical fields.

Such creative travel over the boundary between reform and professionalism proved much more difficult for second-wave feminists because of the success of equal-rights feminism. In the late twentieth century, equal-rights feminism won women job opportunities in mainstream engineering, but because this approach claimed that women were the same as men, the potential opportunities for innovation provided by feminist theory in general and difference feminism in particular were difficult to realize. Women fought for and won job opportunities in engineering, and as they found themselves less marginalized, they had less need for the kind of special career opportunities for women that reform had provided. On the other hand, women who followed a strategy of equal-rights feminism could not use reform commitments or being different or feminist theory as an asset on the job and a source of creativity. Perhaps the clustering of women in certain fields of engineering will make that possible in the future in the way that the clustering of women in some fields of science (such as primatology) has enabled the integration of feminist ideas into those fields.

NOTES

I wish to thank Rima Apple, Angela Creager, John Mauer, Londa Schiebinger, Suzanne Sinke, and Joann Ward for very valuable comments on earlier versions of this manuscript.

1. See Pamela Mack, "Women, the Progressive Movement, and Engineering," paper presented at "Writing the Past, Claiming the Future: Women and Gender in Science, Medicine, and Technology," St. Louis University, Oct. 12–15, 2000.

2. I have found evidence for such a muted equal-rights stance particularly in my unpublished research (with Dr. Miriam Levin) on women scientists at Mount Holyoke College. Historians of women in medicine have suggested a shift in the early twentieth century from an argument about women's special contribution to medicine to an emphasis on participating in the mainstream that seems to me to be a quiet equal-rights feminism. See, for example, Regina Markell Morantz-Sanchez, *Sympathy and Science: Women Physicians in American Medicine* (New York: Oxford University Press, 1985). For a discussion of this strategy among pre–World War II women engineers, see Ruth Oldenziel, "Decoding the Silence: Women Engineers and Male Culture in the U.S., 1878–1951," *History and Technology* 4 (1997): 65–95.

3. For one philosophical analysis of the complexities of this idea, see Rita Felski, "The Doxa of Difference," *Signs: Journal of Women in Culture and Society* 23 (1997): 1–22. There is an interesting overview of the argument about essentialism in Ruth Schwartz Cowan, "Technology Is to Science as Female Is to Male: Musings on the History and Character of Our Discipline," *Technology and Culture* 36 (1996): 572–82.

4. My work on the early twentieth century is developed in detail in "Women, the Progressive Movement, and Engineering." See also this volume, chap. 7.

5. This transition deserves much more attention than I can give here. See, for example, Linda Lear, *Rachel Carson: Witness for Nature* (New York: Henry Holt, 1997).

6. Carolyn Merchant, *The Death of Nature: Women, Ecology, and the Scientific Revolution* (New York: Harper and Row, 1980), xvii.

7. Lynn White, Jr., *Machina ex Deo: Essays in the Dynamism of Western Culture* (Cambridge, MA: MIT Press, 1968). For an interesting pulling together of these ideas in new ways, see David F. Noble, *A World without Women: The Christian Clerical Culture of Western Science* (New York: Knopf, 1992).

8. For an introduction to the connections between ecofeminism and religion, see Nancy R. Howell, "Ecofeminism: What One Needs to Know," *Zygon* 32 (1997): 231–42.

9. Carolyn Merchant, *Earthcare: Women and the Environment* (New York: Routledge, 1995), 139–66.

10. Marlene Longenecker, "Women, Ecology, and the Environment: An Introduction," *NWSA Journal* 9 (1997): 1–18.

11. Sally L. Hacker, *"Doing It the Hard Way": Investigations of Gender and Technology* (Boston: Unwin Hyman, 1990), 121–22.

12. Judy Wajcman, *Feminism Confronts Technology* (University Park: Pennsylvania State University Press, 1991), 146.

13. J. Gregg Robinson and Judith S. McIlwee, "Men, Women, and the Culture of Engineering," *Sociological Quarterly* 32 (1991): 403–21, quotation on 406. The authors also published a book: Judith S. McIlwee and J. Gregg Robinson, *Women in Engineering: Gender, Power, and WorkPlace Culture* (Albany: State University of New York Press, 1992).

14. Ruth Carter and Gill Kirkup, "Women in Professional Engineering: The Interaction of Gendered Structure and Values," *Feminist Review* 35 (1990): 92–101, quotation on 96.

15. Some representative references are Sherry Turkle, *The Second Self: Computers and the Human Spirit* (New York: Simon and Schuster, 1984); Joan Greenbaum, "The Head and the Heart: Using Gender Analysis to Study the Social Construction of Computer Systems," *Computers and Society* 20 (1990): 9–17; Ruth Perry and Lisa Greber, "Women and Computers: An Introduction," *Signs: Journal of Women in Culture and Society* 16 (1990): 74–101; Liesbet van Zoonen, "Feminist Theory and Information Technology," *Media, Culture, and Society* 14 (1992): 9–29; Lynn Cherny and Elizabeth Reba Weise, eds., *Wired Women: Gender and New Realities in Cyberspace* (Seattle, WA: Seal Press, 1996); Nina Lykke and Rosi Braidotti, *Between Monsters, Goddesses, and Cyborgs: Feminist Confrontations with Science, Medicine, and Cyberspace* (New York: St. Martin's Press, 1996); Thelma Estrin, "Women's Studies and Computer Sciences, Their Intersection," *IEEE Annals of the History of Computing* 18, no. 3 (1996): 43–46 (discussed in this volume, chap. 9); Laura DiDio, "Booth Bimbo Bingo," *Computerworld* 46 (Nov. 17, 1997): 98.

16. Representative references for reproductive technologies include Michelle Stanworth, *Reproductive Technologies: Gender, Motherhood, and Medicine* (Minneapolis: University of Minnesota Press, 1987); Patricia Spallone and Deborah Lynn Steinberg, *Made to Order: The Myth of Reproductive and Genetic Progress* (Oxford and New York: Pergamon Press, 1987); Kathryn Strother Ratcliffe, ed., *Healing Technology: Feminist Perspectives* (Ann Arbor: University of Michigan Press, 1989); Helen Bequaert Homes and Laura M. Purdy, eds., *Feminist Perspectives in Medical Ethics* (Bloomington: University of Indiana Press, 1992); Robyn Rowland, *Living Laboratories: Women and Reproductive Technologies* (Bloomington: Indiana University Press, 1992); Susan M. Wolf, *Feminism and Bioethics: Beyond Reproduction* (New York: Oxford University Press, 1996); Anne Donchin and Laura Martha Purdy, *Embodying Bioethics: Recent Feminist Advances* (Lanham, MD: Rowman and Littlefield, 1998).

17. Samuel C. Florman, *The Existential Pleasures of Engineering* (New York: St. Martin's Press, 1976).

18. Samuel C. Florman, *The Civilized Engineer* (New York: St. Martin's Press, 1987), 215.

19. Ibid., 216. Some of the material in Florman's chapter "The Civilized Engineer: Women" was first published in Samuel C. Florman, "Will Women Engineers Make a Difference," *Technology Review* 87 (Nov.–Dec. 1984): 51–52.

20. Others have also suggested that a broader engineering education would both attract more women and improve engineering; see, for example, Sharon Beder, "Towards a More Representative Engineering Education," *International Journal of Applied Engineering Education* 5, no. 2 (1989): 173–82, http://www.uow.edu.au/arts/sts/sbeder/education2.html (Jan. 18, 2001).

21. For example, Martha W. Gilliland, "The Special Voices of Women in Engineering," *SWE: Magazine of the Society of Women Engineers* 40 (July/Aug. 1994): 24–27.

22. One study shows that in management, where such rhetoric has been even more common, women who get ahead tend to use a more traditional style. Judy Wajcman, "Desperately Seeking Differences: Is Management Style Gendered?" *British Journal of Industrial Relations* 34 (1996): 333–50.

23. Carter and Kirkup, "Women in Professional Engineering," 97.

24. Knut H. Sørensen, "Towards a Feminized Technology? Gendered Values in the Construction of Technology," *Social Studies of Science* 22 (1992): 2–31, quotation on 22.

25. Some suggestive U.S. data are available in Kathryn W. Linden, William K. LeBold, Kevin D. Shell, and Carolyn M. Jagacinski, "Perceived Interests of Student and Professional Engineers," *U.S. Woman Engineer* 32 (May/June 1986): 21–24. My own study had a response rate of 73 out of 99 students in a history class for freshman engineers. Of the 12 women, 6 (50 percent) already thought that social issues were important to engineering (before beginning the course), and 6 (50 percent) said they learned from the course that social issues were more important to engineering than they had realized. Of 61 men, 13 (21 percent) said they already thought social issues important, 40 (66 percent) said they learned that social issues were important, and 8 (13 percent) said that engineers should worry about the technical issues and leave the social issues to someone else. The correlation is significant at the .05 percent level.

26. Wajcman, "Desperately Seeking Differences."

27. For example, Hugh Aldersey-Williams, "Want Your Machine to Work? Get a Woman to Design It," *New Statesman* 126 (Aug. 8, 1997): 44–46; Lael Parrot, "Women and the Culture of Engineering: Society Could Benefit from More Female Engineers," *Resource: Engineering and Technology for a Sustainable World* 5 (Jan. 1998): 6–9.

28. Oldenziel, "Decoding the Silence." Oldenziel traces the complex interactions between masculinity and professionalization in engineering in more detail in Ruth Oldenziel, *Making Technology Masculine: Men, Women, and Modern Machines in America, 1870–1945* (Amsterdam: Amsterdam University Press, 1999).

29. Oldenziel, "Decoding the Silence."

30. See ibid.; Alice C. Goff, *Women Can Be Engineers* (Youngstown, OH: 1946); Martha Moore Trescott, "Women in the Intellectual Development of Engineering: A Study in Persistence and Systems Thought," in *Women of Science: Righting the Record,* ed. G. Kass-Simon and Patricia Farnes (Bloomington: Indiana University Press, 1990), 168–76.

31. Mack, "Women, the Progressive Movement, and Engineering."

32. Margaret W. Rossiter, *Women Scientists in America: Before Affirmative Action, 1940–1972* (Baltimore, MD: Johns Hopkins University Press, 1995), esp. 14–16, 55–56.

33. A survey of women engineers by Martha Trescott found that 40 percent of her sample got their first engineering-related job as an intern, trainee, or aide

or in drafting. Martha M. Trescott, "Women Engineers in American History: A Progress Report," *1981 ASEE Annual Conference Proceedings* (Washington, DC: American Society for Engineering Education, 1981), 1123–28.

34. U.S. Bureau of the Census, *Statistical Abstracts for the United States: 1958* (Washington, DC: U.S. Government Printing Office, 1958), 220.

35. Arminta Harness, speaking at the luncheon program, in "Women in Engineering: Conference Proceedings," College of Engineering, University of Washington, Nov. 4, 1974, 18. Another story of entering engineering after the war can be found in Theresa Snyder, "An Interview with Margaret Eller," *U.S. Woman Engineer* 39 (Mar./Apr. 1993): 19–21.

36. Herb Rosen, "Aerospace Pioneer: Katharine Stinson," *U.S. Woman Engineer* 39 (Mar./Apr. 1993): 14–15.

37. Rossiter, *Women Scientists in America,* 236, 259, 278–79, 340. However, the figure for industry also includes self-employed, and it may be that larger numbers of women in that category throw off the figures. It is interesting that women were a higher percentage of engineers employed in industry than of engineering bachelor's degrees (0.3 percent in 1955); this may reflect distortions in the statistics or it may reflect women with degrees in science who took jobs in engineering.

38. Ibid., 339. For more on the history of the Society of Women Engineers, see Alexis C. Swoboda, "The SWE Story," *U.S. Woman Engineer* 39 (Mar./Apr. 1993): 8–12.

39. For a discussion of the importance of considering both changes in raw number and changes in percentage, see Mariam K. Chamberlain, ed., *Women in Academe: Progress and Prospects* (New York: Russell Sage Foundation, 1988), 200–207. One example of quotas is that MIT limited women to 4 percent of the entering class in 1967. Debra Cash, "Mildred Dresselhaus: A Career of Firsts and Onlys," *U.S. Woman Engineer* 39 (Mar./Apr. 1993): 16–18.

40. Leslie A. Barber, "U.S. Women in Science and Engineering, 1960–1990: Progress toward Equity?" *Journal of Higher Education* 66 (1995): 213–35. Colleges and universities were exempt from the Civil Rights Act. The one field in which women found significant new opportunities was computer science, apparently because men did not realize that programming would be a significant and interesting endeavor. See, for example, *Women in Computing,* special issue of *IEEE Annals of the History of Computing* 18, no. 3 (fall 1996).

41. Jessie Bernard, "The Present Situation in the Academic World of Women Trained in Engineering," in *Women and the Scientific Professions,* ed. Jacquelyn A. Mattfeld and Carol G. Van Aken (Cambridge, MA: MIT Press, 1965), 163–82. For an interesting contrasting story of a woman trained in science who moved into engineering, see Mildred S. Dresselhaus, "Electrical Engineer," in *Women and Success: The Anatomy of Achievement,* ed. Ruth B. Kundsin (New York: William Morrow, 1974), 38–43.

42. Amy Sue Bix, " 'Engineeresses' 'Invade' Campus: Four Decades of Debate over Technical Coeducation," in *Women and Technology: Historical, Societal, and Professional Perspectives,* Proceedings of the 1999 IEEE International Symposium on Technology and Society (Piscataway, NJ: IEEE, 1999), 195–201.

43. Rossiter, *Women Scientists in America,* 361. Rossiter documents the influence of feminism on key women scientists but does not consider engineering separately.

44. "Nance Dicciani—A Bias for Action," *U.S. Woman Engineer* 32 (Jan./Feb. 1985): 5.

45. Chamberlain, *Women in Academe,* 220.

46. Affirmative action was first included in an executive order prohibiting discrimination in the early 1960s, and sex discrimination was added to the list of prohibited forms of discrimination in 1968. Chamberlain, *Women in Academe,*

171. The quotation about the demand for women engineers is from a talk by Lynn Colella Bell in "Women in Engineering: Conference Proceedings," College of Engineering, University of Washington, Nov. 4, 1974, 14.

47. Jane Zimmer Daniels, "Organizations, Conferences, Publications, and Funding for Women in Engineering: A Historic Review," *SWE: Magazine of the Society of Women Engineers* 41 (Mar./Apr. 1995): 22–26.

48. Rossiter, *Women Scientists in America,* 374–76; Chamberlain, *Women in Academe,* 172–74, 181–83.

49. For a careful definition and discussion of affirmative action, see Chamberlain, *Women in Academe,* chap. 8.

50. I am very serious about what engineering professors learned from their daughters; when I talked about these issues with senior engineering faculty ten years ago, I frequently heard stories of their daughters' professional struggles (not necessarily in engineering). However, the fatherly attitude could also be a form of subtle sexism: "So long as women are young and vulnerable, competing only for subordinate or junior positions, men relate comfortably with them" (Taggart Smith, "Women Faculty in an Engineering Environment: A New Look," *U.S. Woman Engineer* 38 [Mar./Apr. 1992]: 27).

51. I write this paragraph in part from my own experience as a science student (Harvard, 1977) and my impressions of the experience of a woman friend who was an engineering major. For the chilly classroom climate, see Roberta M. Hall, *The Classroom Climate: A Chilly One for Women,* Project on the Status of Education of Women (Washington, DC: Association of American Colleges, 1982). For examples of the chilly climate at engineering schools, see Bix, "'Engineeresses' 'Invade' Campus."

52. Linda Geppert, "The Uphill Struggle: No Rose Garden for Women in Engineering," *IEEE Spectrum* 32 (May 1995): 44.

53. My conceptualization of this comes from how I and other young women in astronomy in the mid-1970s understood the attitudes of some of the older women astronomers. One description of this pattern in engineering describes women engineers' earlier "unwillingness to be associated with other women. Some women in engineering, as in other disciplines, used to say, 'I think of myself as a scientist not as a woman scientist'" (Jennie Farley, "Women Getting Ahead in Engineering," *U.S. Woman Engineer* 36 [Sept./Oct. 1990]: 16).

54. Theresa Snyder, "An Interview with Margaret Eller," *U.S. Woman Engineer* 39 (Mar./Apr. 1993): 19–21. Eller was born in 1913 and worked in drafting and engineering education from World War II to 1988.

55. Anil Saigal, "Women Engineers: An Insight into Their Problems," *Engineering Education* 78 (1987): 194–95.

56. Geppert, "The Uphill Struggle," 40–50, esp. 47–49. Stephen G. Brush, "Women in Science and Engineering," *American Scientist* 79 (1991): 404–19. The idea that discrimination is no longer a serious problem may be false; in a 1993 survey by the Society of Women Engineers, 58 percent of the women who filled out the survey said they were personally aware of instances of discrimination. Patricia L. Eng, "Who Are We? How Different Are Men and Women Engineers?" *Today's Engineer* (winter 1998): 40–43.

57. Geppert, "The Uphill Struggle," 42–43. Changes in automobile technology changed this practice; many fewer young men came to school in the 1990s with experience tinkering with automobile engines.

58. There are some examples of this in Laura Shefler, "A Picture of Change," *U.S. Woman Engineer* 39 (Mar./Apr. 1993): 30–34.

59. See the webpages of these organizations for examples. Women in Engineering: Program Advocates Network is at http://www.engr.washington.edu/~wepan, The Society for Women Engineers is at http://www.swe.org, and

IEEE Women in Engineering is at http://www.ieee.org/organizations/committee/women/homepg.htm (Jan. 18, 2001). The National Academy of Engineering's Celebration of Women in Engineering is at http://www.nae.edu/nae/cwe/cwe.nsf/Homepage/NAE+Celebration+of+Women+in+Engineering?OpenDocument (Jan. 18, 2001).

60. Elizabeth Culotta, Patricia Kahn, Toomas Koppel, and Ann Gibbons, "Women Struggle to Crack the Code of Corporate Culture," *Science* 260 (Apr. 16, 1993): 389–404.

61. Robinson and McIlwee, "Men, Women, and the Culture of Engineering," 406.

62. Ibid., 417.

63. Parrot, "Women and the Culture of Engineering," 7.

64. Culotta, Kahn, Koppel, and Gibbons, "Women Struggle to Crack the Code of Corporate Culture."

65. Adele Scheele, "Learning to Compete with Women," *U.S. Woman Engineer* 39 (Mar./Apr. 1993): 38–39.

66. Barber, "U.S. Women in Science and Engineering," 232. Unfortunately, Barber's discussion of this conclusion cites examples only from science. One organization that seems to be an exception to the pattern of not trying to change the system is Catalyst, a consulting company working to advance women in business. They published a report in 1992, "Women in Engineering: An Untapped Resource," that discusses how women engineers are subject to "intense pressure to conform to 'masculine' styles." For more information see http://www.catalystwomen.org (Oct. 8, 2000).

67. Trescott, "Women Engineers in American History."

68. Alan Elzerman, conversation with author, Aug. 18, 1998.

69. Figures from National Science Foundation, *Women, Minorities, and Persons with Disabilities in Science and Engineering: 1996,* NSF 96-311 (Washington, DC, Sept. 1996), appendix table 3-26.

70. Monica Frize, "Managing Diversity," paper written for the 1998 "More than Just Numbers Conference" in Vancouver and available at http://www.carleton.ca/wise/mtjn98pap.html (Oct. 8, 2000).

71. A group of women engineers talk frankly about this in Geppert, "The Uphill Struggle," 44–46.

NINE

Boys' Toys and Women's Work: Feminism Engages Software

MICHAEL S. MAHONEY

In her contribution to this volume, Pamela Mack reflects on her inability to find a case in which feminism, particularly feminist theory, has improved practice or made for better engineering. What follows is an extended version of a commentary meant at the time to take up her implicit challenge by considering a future case study of a situation still in flux, one in which feminism might still make a difference in ways we can now specify. The field in question is software engineering, but before turning to it I would like to consider the computer and computing more generally as a case study of gender and technology.

As a case study, computing has the advantage of having emerged as an essentially new technology at a definable time and place. Before the creation of the electronic digital computer in the mid-1940s, what we now understand as computing did not exist. Before then, "computer" denoted a person, usually a woman, who carried out calculations by hand or with a mechanical calculator. The work was viewed as clerical: computers followed explicit instructions (usually provided by a male mathematician), and accuracy took precedence over imagination. Indeed, when Turing wanted an example of a "mechanical" procedure, he invoked the human computer, albeit in masculine form.

Perhaps for that reason, the first programmers for the ENIAC were women,[1] as were several of the leading figures in the early development of computing, most notably Grace Hopper. Yet, men soon took over, and leading women became the exception in an increasingly masculine field. Hopper remained highly visible until her death in 1992 at age eighty-five. Indeed, "Amazing Grace" achieved something akin to canonization in her own lifetime.[2] Everyone of a certain generation and older has a Hopper tale to tell, and the telling often has a hagiographical tone. Although Jean Sammet did not reach that exalted state of matron saint, she also remained a formidable figure within IBM and the profession, universally recognized for her encyclopedic knowledge of programming languages. One can name others, several of whom have served as presidents of the Association of Computing Machinery (ACM), but that is perhaps the point: one can name them.

How and why did the field fall to men? Does the answer go beyond the obvious, namely that in computing, as in other fields of war work, women returned to the home, or were directed back to the home, to make room for men returning to the labor force? The women of EN-IAC, the first programmers, tell a common story of careers interrupted or ended by marriage and the arrival of children.[3] Or is it a matter of men discovering that programming, unlike calculating, was a challenging, creative intellectual enterprise that promised rewards and reputation? At one of the first conferences on the history of computing, John Backus described the culture of the early programming community:

> Programming in the America of the 1950s had a vital frontier enthusiasm virtually untainted by either the scholarship or the stuffiness of academia. The programmer-inventors of the early 1950s were too impatient to hoard an idea until it could be fully developed and a paper written. They wanted to convince others. Action, progress, and outdoing one's rivals were more important than mere authorship of a paper. Recognition in the small programming fraternity was more likely to be accorded for a colorful personality, an extraordinary feat of coding, or the ability to hold a lot of liquor well than it was for an intellectual insight. Ideas flowed freely along with the liquor at innumerable meetings, as well as in sober private discussions and informally distributed papers.[4]

In short, programming quickly became a hard-drinking boys' club; the use of "fraternity" is revealing. By the late 1950s, men appear to have had the field to themselves. The groups that gathered informally at the RAND Corporation a day prior to the annual Western Joint Computer Conference to discuss the state of the profession included no women.

THE (TRANS)GENDERING OF THE COMPUTER

As the field became masculine, so too did the machine. At first, that may not seem puzzling. Given the traditional gendering of machines, what else would one expect of a room full of vacuum tubes and oscilloscopes, even as the tubes disappeared behind metal doors and the oscilloscope gave way to the video display? Nevertheless, two questions immediately arise. First, by the early 1950s programming was emerging as an activity distinct from that concerned with the computer as a physical device. The split was reinforced with the development in the mid- to late 1950s of programming languages and operating systems, both designed to keep the programmer away from the machine.[5] In most computing installations, the computer had a room of its own, in which it was tended by designated operators, often women, who mediated between it and the programmers. Academically and professionally, computer engineering took charge of the hardware, while computer science concerned itself with the software, that is, with programming languages, operating systems, and methods of data management. Hence, although continuing cultural icons might explain the masculine character of the computer itself, they do not explain why programming should have been masculinized. The boys' club is not an explanation; it needs explaining.

Second, the personal computer assumed a shape in the late 1970s and early 1980s that tied it to cultural icons and activities that in and of themselves seem less straightforwardly masculine. The video screen, keyboard, and mouse invited users to write, to draw, to play games, to communicate. Indeed, the emphasis on the verbal, the visual, and the conversational would seem to have pushed the device more toward the feminine, but apparently, that is not how the culture has viewed it. Indeed, casting the computer in a quite different form does not seem to have changed things that much. Studies since the early 1980s have generally found that children of both sexes from kindergarten on identify the personal computer as masculine: it is something for the boys.[6] The researchers themselves have found that puzzling, but it may at least help to explain why the fundamental restructuring of the computer industry reflected in the replacement of IBM by Microsoft as the shaping force has apparently done little to alter cultural perceptions of the technology. The "triumph of the nerds" does not seem a victory for women.

The central question seems clear. Computing is readily available public knowledge: anyone can acquire it. The machine is for sale everywhere: anyone can buy it. Any woman who wishes to do so can become a computer scientist, computer engineer, or software developer.[7] Yet, by any measure, the world of computing includes relatively few women. The reason seems at first equally clear. The door may be open, but the world beyond it does not invite entry. Computing is a masculine world, in which women do not feel comfortable. However, that is a restatement of the question, not an answer to it. It does not explain what it means for computing to be masculine and how it became so.

Nor does it suggest what a regendering of the world of computing might entail and how it might be accomplished. By the same token, it does offer a first-order reply to the question that shapes this volume: what difference has feminism made? To judge by the literature on women and computing, none. But that is a first-order reply, and it requires some adjustment.

Indeed, when one examines the question closely, it fragments into pieces that grow more puzzling as they become more specific. Are we talking about computer science or computer technology? In the latter case, are we talking about the technology of production or about computers out there in "consumption junction"?[8] It is evident that women have not played, and still do not play, much of a role in computer science or in the design and development of computer systems, whether mainframe, minicomputer, or personal computer. They do seem to be more proportionately represented among the consumers of computing, where their representation would seem to be determined more by their presence in business and the professions than by a particular stance toward computers. But there, too, computing seems to undermine that presence by making women and their work invisible. All that seems clear. The problem is to explain why and to do so without naturalizing technology or essentializing gender.[9]

Computing lends itself to the challenge, since the computer as a concept is protean. In principle, a computer can do anything we can describe in certain ways that are limited more in theory than in practice. So in practice computing is what we have made of it since the creation of the computer in the late 1940s and early 1950s.[10] Whatever one wants to say about such abstractions as the Turing machine, it is hard to know how physical computers and the systems running on them could be anything other than socially constructed.[11] Computing has no nature. It is what it is because people have made it so.

Those people have been overwhelmingly men. That is a readily verifiable fact. What they have created is overwhelmingly masculine. That is a readily verifiable perception, which one often reads in feminist explanations of the paucity of women in the field. Yet, although writers on the subject of gender and computing all proceed on the premise that computing is gendered masculine—it must be, else why would it be dominated by men?—the more closely they look at it, the less clear it is just what makes it so. As Judy Wajcman and others have pointed out, some of those explanations verge on the essentialist, as "masculine" and "feminine" take the pole positions in a set of dichotomies: abstract/concrete, objectivity/subjectivity, logical/intuitive, mind/body, dominating/submissive, war/peace, and so on.[12] Computing generally winds up on the masculine side of each of them. But essentialism is not the only problem here. Under such an interpretation, a feminine form of computing becomes the complete obverse of what now exists, and that seems as unimaginable as it is unrealistic.[13] Fortunately, a closer look, both at the dichotomies and at their application to computing, suggests that the solution is not that drastic. To show how, we need first to consider the ways in which efforts to pin down the masculine nature of computing end up blurring the very lines they try to draw. Let us work with three case studies from the recent literature.

CROSSING BOUNDARIES

In "Women's Studies and Computer Science: Their Intersection,"[14] Thelma Estrin points to convergences between feminist epistemology on the one hand and developments in computing and cognitive science on the other. Aimed at reconciling two disciplines that "both emerged as academic disciplines in the 1960s, with no interaction between them because they evolved along very different paths—one for women and one for men,"[15] her discussion tends to cloud the grounds of their convergence rather than shedding light on them. For, without evident input from women or feminism, a masculine computer science seems by Estrin's own description to have developed precisely the modes of thinking that she and others want to identify as feminine.

For example, asserting that "[f]eminist epistemology supports concrete thinking as a valuable tool in our way of thinking,"[16] Estrin makes much of object-oriented programming as a style of thinking in concrete terms. Contrasted with procedural and functional programming, "object-oriented programming is regarded as a physical model simulating the behavior of either a real or imaginary part of the world (C++). Physical models have elements that directly reflect phenomena and concepts that undermine the canonical position by supporting trends that challenge established methods and encourage working with specific objects."[17] The "(C++)" refers to perhaps the most popular object-oriented programming language in use today, which Estrin interestingly contrasts with C and with Lisp as examples of the other modes of programming. The contrast is interesting for two reasons. First, C++ began life as a preprocessor for C; it was originally "C with Classes" and based on Simula, a simulation language dating back to the 1960s. It was created in exactly the same environment as C, namely the Computing Research department at Bell Labs, Murray Hill.[18] Second, Lisp was initially designed to facilitate just the sort of interactive programming that Estrin (following Sherry Turkle and Seymour Papert) identifies with "bricolage," and it was the basis for Logo, the children's language that Estrin praises highly for encouraging thinking about objects and their relations. Both Lisp and Logo are products of MIT's AI Laboratory, not particularly renowned for its feminist outlook.[19] Moreover, Lisp today comes in flavors that include objects, which are conceptually quite compatible with it.

If one considers the history of object-oriented programming in terms of its genealogy, then its proposed empathy with feminist modes of thinking appears all the more problematic. Estrin points specifically to Smalltalk, the language of Alan Kay's Dynabook, aimed at placing the power of the computer in the hands of children for drawing, writing, and creative programming.[20] But, like C++, Smalltalk itself is also an offspring of Simula, which was designed not for children but for researchers doing systems analysis and simulations at the Norwegian Computing Center in the early 1960s.[21] Within that genealogy, object-oriented programming emerges from the notion of modularity, or the division of a program into subunits that are independent of one another and of any specific context. An early and well-known expression of the notion is M. D. McIlroy's proposal in 1968 for "mass-produced software components," which is filled with analogies to machine tools and redolent with the smell of the machine shop and indeed the assembly line.[22]

Although one might think of object-oriented programming as rooted in the concrete, object-oriented design involves a sustained process of abstraction, in which one seeks to characterize the objects as members of classes, which themselves are elements of yet more general classes, so that objects may inherit the properties they share with members of other classes. Thus the object "square" is a kind of "rectangle," which in turn is a kind of "figure," and so on. The result is a hierarchy of classes, another concept not generally associated with the feminine. Deciding on what will be the objects constituting a program and how they are related with one another seems a quintessentially Aristotelian exercise in definition by abstraction. To the extent that abstract, hierarchical thinking is classified as masculine, it is difficult to see how object-oriented design and programming are any less so.

If, then, there is convergence between computer science and women's studies, it does not seem to result from the reshaping of computational thinking by feminist epistemology. What Estrin identifies as common elements have emerged within computing from what would seem to be masculine foundations. If the elements to which she points open computer science to women by making it amenable to their ways of thinking about the world, then it has been open for a long time, at least back to the 1960s, because they are ways of thinking to which the men who created computing themselves aspired.

Tove Håpnes and Knut Sørensen's study of a Norwegian hacker culture similarly, but in their case intentionally, undermines what earlier seemed certainties about the ways in which computing is masculine. "One of the reasons we became interested in this group was the conclusions from a study of female computer science students. They used the hackers as a metaphor for all the things they did not like about computer science: the style of work, the infatuation with computers leading to neglect of normal non-study relations, and the concentration on problems with no obvious relation to the outside world. Thus, the hackers emerged as a possibly important example of an extremely masculine technological culture."[23] Contrary to expectation, Håpnes and Sørensen encountered practices and attitudes that straddle gender lines: the hackers turned out to be competitive but communal and mutually supportive, "hard masters" yet open to the approaches characteristic of "soft mastery," fascinated by the machine yet concerned to write useful programs. The findings both confirm and confound the pictures painted by Joseph Weizenbaum and Sherry Turkle, touchstones for assays of this sort.[24] "Compared to Turkle's description of American MIT hackers," they write, ". . . the Norwegian hackers appear as less extreme and more 'feminine.' "[25] But, they go on, that might have something to do with differences in the national cultures.

Perhaps it does, but then we would have a curious separation of culture from its material base. Something is missing here. Although Håpnes and Sørensen refer at points to specific technical practices, they do not explore the role of the programming environment in enabling or even encouraging the behavior and attitudes they observed. They note that the hackers eschew Pascal and COBOL in favor of C and that they belittle the Macintosh while touting the virtues of their Sun workstations. But it is quite likely not the workstation alone that they are praising, because C does not stand by itself on that platform. It seems likely that the hackers are programming in the Unix environment, which was quite consciously designed to foster certain patterns of behavior both individually and collectively. Since the turn of the 1980s, it has been the dominant operating and programming system in academic computing and, as a model at least, in the microcomputer software industry. Indeed, in the form of Linux it has recently begun to eat away at the near-monopoly of Microsoft Windows.

Anyone familiar with Unix will recognize it in the behavior of the Norwegian hackers. Unix began as a system for sharing files and thus for communal work.[26] Although Ken Thompson and Dennis Ritchie provided the basic system and the C language, respectively, the development of Unix was a collaborative effort among a relatively large number of people both at Bell Labs and later at Berkeley. The source files were open to all, and anyone could decide to enhance or modify any part of the system. With that freedom came responsibility in accordance with the rule that whoever touched a routine assumed the task of maintaining it for the others until someone else decided to make a change. Membership in the community meant mutual support.

The guiding metaphor of Unix is the toolbox. Unix is an environment for artisans, who craft software from a variety of small programs that "do one thing and do it well" but also can be linked to one another in a sequence, or "pipe," each taking input from its predecessor and supplying output to its successor. Many of these programs constitute "little languages" and emphasize the verbal aspects of computing. Unix thus encourages what developers refer to as "rapid prototyping" but what Turkle and others would immediately recognize as "bricolage." Pipes and the little languages make programming an interactive enterprise, allowing the programmer to piece together solutions to problems before committing to a final program. That is what makes it appealing to Håpnes and Sørensen's hackers, but it is also what ought to make it attractive to Estrin as a programming environment for women, given her view of feminist epistemology.

Unix as a mainstream culture of computing seems to present considerable ambivalence here and thus provides an interesting perspective from which to view, as a third example, Ulrike Erb's "Exploring the Excluded: A Feminist Approach to Opening New Perspectives in Computer Science." Erb undertook an empirical study of "[German] female computer scientists, who, in spite of the hacker culture and women's marginalization, have successfully made their way in computer science." She found that a "techno-centered" culture of computing acted as an obstacle to her subjects, who had sought access to the field by way of theoretical computer science:

> The interviews show that even among computer scientists it remains vague what it means to be competent in computer technology or to be a "technical insider": Does it mean sitting all day and night in front of the computer? Does it mean knowing every bit and byte or knowing the latest software? Or does it mean knowing how to construct blinking surfaces? In the interviews it became evident that in fact all this is associated with technical competence, while on the other hand knowledge on [sic] system design like requirements engineering, specification and design of algorithms is not thought to belong to technical competence.[27]

Erb emphasizes that this was often a matter of perception. In actual practice, the women displayed much of the technical competence they denied possessing, largely because they did not want to be associated with the work habits it connotes. By the same token, rapid prototyping did seem to attract more attention and praise from the profession at large than did more considered specification and design based on inquiry into users' needs and responses, an approach to which the women felt particularly drawn.

One would like to ask how they feel about Unix. On the one hand, the Unix culture rewards technical virtuosity and fosters rapid prototyping and exploratory approaches to design. On the other hand, it reflects fundamental theoretical concerns. Among those who contributed to it in both spirit and substance are Alfred Aho, John Hopcroft, and Jeffrey Ullman, who teamed in several combinations to produce standard texts in computer science, including the theory of automata and formal languages, the design of algorithms, and the theory and design of compilers. Those texts can be found in the computer science section of German, as well as American, bookstores. Other contributors to Unix have written equally accessible standard treatises in computational complexity. Unix is theory based in ways that make it a model for the approach to computing that Erb's women claim to admire and that they see as their means of access to the field. And it is both the operating system and the programming environment in which AT&T maintains its Electronic Switching System, the backbone of the U.S. telephone system, designed very much with the user in mind.

Unix aside, Erb's point quoted above poses another problem for an analysis based on gender. The areas of requirements analysis, specification, and high-level design have formed part of software engineering from its inception in the early 1970s. Since the mid-1980s they have tended to dominate the attention of software engineers, both in attracting its leading practitioners and in forming the focus of what those practitioners consider best practice. No one doubts the crucial role of formal requirements analysis and specification; every textbook in software engineering emphasizes it, and the bulk of the research effort in the field has gone toward the development of tools to facilitate the work. In software engineering as a subdiscipline of computer science, formal methods of requirements analysis, specification, and design have drawn the major share of professional attention and recognition. As far as the theoretical community is concerned, Erb's women view the field in a way that would be difficult to corroborate by more or less objective evidence of its actual practice. So there is a problem of perception here, too.

That said, it is also the case that one of the dilemmas of software engineering from its beginnings is that what the theoretical community deems best practice is seldom reflected in the way that large software projects have actually been carried out. Studies have repeatedly shown that the most serious and most expensive errors occur in the early stages, in particular in setting down clearly what the system is supposed to do. Yet, they also show that only a small fraction of project time is devoted to those early stages. Most commonly, technical proficiency does indeed take the place of thoughtful analysis and design, and the outcome is often disastrous. But, in principle at least, the discipline deplores the problem. To that extent, Erb's subjects are in the mainstream, not bucking the current.

SOFTWARE, ENGINEERING, AND SOFTWARE ENGINEERING

Erb concludes her article by urging that "there is a great need for women's research and in particular for women's research done by female computer scientists. In this way women's studies can help to reveal the excluded and *to integrate the excluded in order to enrich computer science by means of the forgotten perspectives.*"[28] Erb does not specify what perspectives she believes have been ignored or forgotten, and all three articles make it hard to see what they might be, since none of the usual suspects seems on close examination to be missing from the lineup. By contrast, the current state of software engineering suggests how her general call for reform might be made specific.

The research of Lucy Suchman and her colleagues on participatory design is concerned, among other things, with situated cognition, or knowledge and skill that are invoked in specific circumstances. They are often tacit and hence go unnoticed or undervalued in studies of use and practice on which the design of computer systems is based.[29] Those studies focus on what can be formulated in general terms: one does not usually computerize this or that traffic control room, but traffic control rooms in general. For historical and cultural reasons, women's work and their ways of carrying it out quite often disappear from view in this move from the situated to the general, for example, the experience and judgment underlying clerical tasks or the often critical decisions routinely made by nurses but not allocated to them professionally. This is where Erb sees promise in general. If we look at the professed interests of her subjects, that promise assumes a much more definite shape. We need to back up just a bit to see it.

Between 1955 and 1970, through a pattern of interaction that remains largely unanalyzed, the use of computers spread widely while the computer industry expanded. The ever wider variety of applications, combined with the increasing sophistication of computer systems themselves, put an exponentially growing pressure on programmers to keep pace with the demand. By the end of the 1960s, almost every large-scale programming project had fallen behind schedule, exceeded its budget, and failed to meet its specifications. Software was not keeping pace with hardware. There were never enough programmers, yet managers were learning that adding more programmers to a late project only made it later.

The dominant response in the industry was to view the situation as a problem in production, to be solved by some form of engineering. In 1968, that view started on the path to institutionalization with the convening of the first NATO Software Engineering Conference in Garmisch, Germany; a second followed in 1969 in Rome. "The phrase 'software engineering' was deliberately chosen as being provocative, in implying the need for software manufacture to be based on the types of theoretical foundations and practical disciplines that are traditional in the established branches of engineering."[30] Although individual uses of the term "software engineering" predated these conferences, they established it as a field of computing, and in the intervening thirty years it has acquired most of the hallmarks of a professional engineering discipline: societies (or groups within societies), journals, conferences, textbooks, curriculum, and so on. One thing is missing, however, and it nags at practitioners: they cannot agree among themselves that it is in fact an engineering discipline and, if it is not, what would make it so. In the meantime, the problems software engineering was supposed to solve have continued or grown even worse.

One can take several different and revealing directions in exploring this situation. For present purposes, three observations may suffice. First, stating the problem so as to make its solution a form of engineering committed practitioners of the new field to historical and cultural models that have traditionally been associated with men. One model was engineering as applied science, which led to sustained efforts to reduce software development to mathematically based programming tools. Another model was industrial engineering, which fostered the notion of the "software factory" along both Taylorist and Fordist lines. Closely allied to the industrial model was the view of engineering as project management, also with its Taylorist forms of organization of production by chain of command. Other models were open at the time, among them architecture and craft practice, but they received little attention.

Second, no woman participated in either conference, although several women have subsequently established a strong presence in the field, precisely in the areas preferred by Erb's women, namely requirements analysis, specification, and design.[31] Interestingly, in her account they construe those areas as belonging to computer science, that is, as pertaining to theoretical analysis rather than technical experimentation. Theory, to repeat one of Erb's main findings, provides them with access to a world of computing that they otherwise view as uninviting or even hostile.

That leads to the third and concluding point, which a diagram will help to make (fig. 9.1). The goal of software development is to model a portion of the real world on the computer. The process begins with the analysis and specification of just what one wants to model and how the model is to behave. It continues with the design and coding of that specification, which results in a working program that in essence is what is known as a finite-state machine.

From the very beginning, software engineers have known through evidence and experience that most of the errors and deficiencies of systems result from inaccurate, inconsistent, and incomplete specification of the job to be accomplished. Those problems are the most difficult and most expensive to solve. That is why so much research effort has gone into developing formal methods and tools for requirements analysis and specification. People in the field have sought nothing less than mathematical completeness and verification of the model to be built. That is, they have striven to achieve for the upper half of the process in the diagram what computer science in the 1960s and 1970s accomplished for the lower half: mathematical certainty.[32] Very few errors creep in during coding, not least because the software tools include powerful diagnostics.

One may question, however, whether the upper half of the process has anything to do with computer science or with software engineering on any of its (now) traditional models. It begins, to repeat, with a determination of what is to be modeled, and that involves an understanding not of computers but of the real-world situation in question. It means in particular an understanding of what the people in that situation are trying to accomplish, what they contribute to accomplishing it, and how a computer system would help them to do it better or at least more efficiently. That is not what one learns in studying computer science; that is not what computer science is about.[33]

To accomplish that task requires that one know how to observe and to listen. To take a lead from Suchman, it requires being able to see what certain perspectives render invisible, to hear what certain

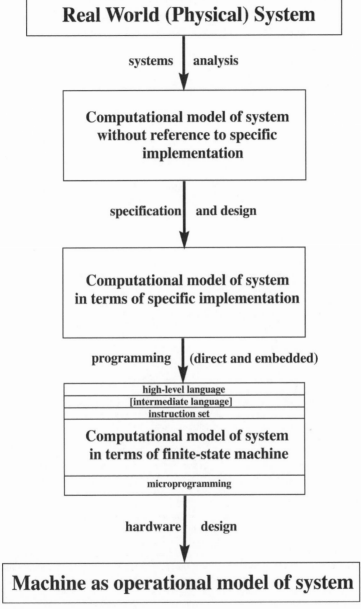

Figure 9.1 Stages of software development

discourses render inaudible. It places a premium on asking how the system is currently working before dictating how it should work in the future. Here, it seems to me, is where feminism could make a difference to one field of engineering (whether or not it is in fact engineering). Feminist analysis has brought out the ways in which a world built by men hides the ways in which women make it work, and it is the working world that computers must capture if they are going to enhance all our lives.

NOTES

1. The Electronic Numerical Integrator and Calculator (ENIAC) went into operation on February 14, 1946, at the University of Pennsylvania's Moore School of Engineering. It was the basis for the subsequent development of the stored-program device now commonly understood as the electronic computer.

2. Her name lives on in particular in the Grace Murray Hopper Award of the Association of Computing Machinery (for outstanding achievement by a researcher under thirty; established 1971, not yet awarded to a woman) and in the destroyer USS *Hopper,* commissioned in 1997 (http://www.hopper.navy.mil).

3. W. Barkley Fritz, "The Women of ENIAC," *IEEE Annals of the History of Computing* 18, no. 3 (1996): 13–28; and Jennifer Light, "When Computers Were Women," *Technology and Culture* 40, no. 3 (1999): 455–83. On the loss of technical jobs following the war, see Margaret W. Rossiter, *Women Scientists in America: Before Affirmative Action, 1940–1972* (Baltimore, MD: Johns Hopkins University Press, 1995).

4. John Backus, "Programming in America in the 1950s—Some Personal Impressions," in *A History of Computing in the Twentieth Century,* ed. N. Metropolis et al. (New York: Academic Press, 1980), 126–27.

5. See Michael S. Mahoney, "Software: The Self-Programming Machine," in *Creating Modern Computing,* ed. Atsushi Akera and Frederik Nebeker (New York: Oxford University Press, forthcoming).

6. For the first round of studies, see *Women, Girls, and Computers,* special issue of *Sex Roles* 13, no. 3/4 (1985), which includes empirical studies ranging from kindergarten to adults in the office. Subsequent studies have in general reinforced the basic tendency, but with some conflicting findings. For a critical evaluation of seventy reports, see Robin Kay, "An Analysis of Methods Used to Examine Gender Differences in Computer-Related Behavior," *Journal of Educational Computing Research* 8, no. 3 (1992): 277–90; and for a recent effort to disaggregate the factors involved, see Lori J. Nelson and Joel Cooper, "Gender Differences in Children's Reactions to Success and Failure with Computers," *Computers in Human Behavior* 13, no. 2 (1997): 247–67. Despite the conflicts and critiques, the general trend of these results would seem to have profound implications for educational policy, but few commentators, much less school boards, seem to have considered the effects of introducing into classrooms devices that students so clearly associate with one gender rather than the other. Interestingly, the question did not appear to have caught the attention of the Clinton administration in Washington, perhaps because it placed its concern for women's issues in conflict with its enthusiasm for educational technology.

7. Perhaps at this point, I should clarify my perspective. My daughter majored in computer science and music at a research-oriented university and has had a successful career as a software developer. Her experiences as a woman in

computing have been a continuing topic of conversation between us for some fifteen years.

8. Ruth Schwartz Cowan, "The Consumption Junction: A Proposal for Research Strategies in the Sociology of Technology," in *The Social Construction of Technological Systems: New Directions in the Sociology and History of Technology*, ed. Wiebe E. Bijker, Thomas P. Hughes, and Trevor J. Pinch (Cambridge, MA: MIT Press, 1987), 261–80.

9. Flis Henwood, "Establishing Gender Perspectives on Information Technology: Problems, Issues and Opportunities," in *Gendered by Design? Information Technology and Office Systems,* ed. Eileen Green, Jenny Owen, and Den Pain (London and Washington, DC: Taylor and Francis, 1993), chap. 2.

10. "Computer" here means electronic, digital, stored-program computer, that is, a practical device with the capabilities of a Turing machine. One can get bogged down in various definitional problems here, none of which is pertinent to the point at issue.

11. That does not mean anything goes. No one who has worked with computers doubts their capacity to resist (see Andrew Pickering, *The Mangle of Practice* [Chicago: University of Chicago Press, 1995]). That resistance goes beyond occasional crashes in the midst of a late-night chapter: it includes planes crashing, rockets going astray and exploding, overdoses of radiation therapy, and collapse of the telephone system. These seem about as real and nonnegotiable as the world can get.

12. Judy Wajcman, *Feminism Confronts Technology* (University Park: Pennsylvania State University Press, 1991).

13. Or, to take another tack, counterproductive: "It could equally well be that once these newly-discovered [feminine] attributes of flexibility, intuition, etc. are revalued and become sought-after skills in computing, men will be the first in line to demonstrate their competence in the field" (Henwood, "Establishing Gender Perspectives," 42–43).

14. Thelma Estrin, "Women's Studies and Computer Science: Their Intersection," *IEEE Annals of the History of Computing* 18, no. 3 (1996): 43–46. Estrin is professor emerita of computer science at the University of California at Los Angeles and has had a distinguished career in the field of biomedical engineering and computer science, including terms as director of the Engineering, Computer, and Systems Division of the National Science Foundation, 1982–84, and member of the Board of Directors of the Aerospace Corporation.

15. Ibid., 43.

16. Ibid., 44.

17. Ibid., 46.

18. Bjarne Stroustrup, *The Design and Evolution of C++* (Reading, MA: Addison-Wesley Publishing Co., 1994).

19. By coincidence Estrin's article is followed in the same special number of *Annals* by Alison Adam's "Constructions of Gender in the History of Artificial Intelligence," in which the line of thinking for which Lisp has served as primary tool, indeed for which it was designed, is designated as irremediably masculinist. It is hard to aim at Lisp without hitting Logo.

20. Alan Kay, "The Early History of Smalltalk," in *History of Programming Languages II,* ed. T. M. Bergin and R. G. Gibson (New York: ACM Press; Reading, MA: Addison-Wesley, 1996), 511–78.

21. Kristen Nygaard and Ole-Johan Dahl, "The Development of the Simula Languages," in *History of Programming Languages,* ed. Richard Wexelblat (New York: Academic Press, 1981), 439–80.

22. M. Douglas McIlroy, "Mass Produced Software Components," in *Software Engineering: Report on a Conference Sponsored by the NATO Science Committee, Garmisch, Germany, 7th to 11th October 1968,* ed. Peter Naur and

184   MICHAEL S. MAHONEY

Brian Randell (Brussels: NATO Scientific Affairs Division, 1969), 138–50; reprinted in *Software Engineering: Concepts and Techniques, Proceedings of the NATO Conferences*, ed. Peter Naur, Brian Randell, and J. N. Buxton (New York: Petrocelli/Charter, 1976).

23. Tove Håpnes and Knut Sørensen, "Competition and Collaboration in Male Shaping of Computing: A Study of a Norwegian Hacker Culture," in *The Gender-Technology Relation: Contemporary Theory and Research*, ed. Keith Grint and Rosalind Gill (London and Bristol, PA: Taylor and Francis, 1995), 177. They refer here to B. Rasmussen and T. Håpnes, "The Production of Male Power in Computer Science," in *Women, Work, and Computerization*, ed. I. V. Erikson et al. (Amsterdam: North-Holland, 1991); and T. Håpnes and B. Rasmussen, "Excluding Women from the Technologies of the Future?" *Futures* 23, no. 10 (1991): 1107–19.

24. Joseph Weizenbaum, *Computer Power and Human Reason* (San Francisco: W. H. Freeman, 1976), esp. chap. 4, "Science and the Compulsive Programmer"; Sherry Turkle, *The Second Self: Computers and the Human Spirit* (New York: Simon and Schuster, 1984), esp. chap. 3, "Child Programmers: The First Generation." In discussing styles of programming, Turkle differentiates between hard and soft mastery, which she associates with Claude Lévi-Strauss's distinction between scientist and *bricoleur*: "Hard mastery is the mastery of the planner, the engineer, soft mastery is the mastery of the artist: try this, wait for a response, try something else, let the overall shape emerge from an interaction with the medium. It is more like a conversation than a monologue" (104–5). Although her first example contrasts the practices of two boys, she goes on to note, "But now it is time to state what might be anticipated by many readers: girls tend to be soft masters, while the hard masters are overwhelmingly male" (109).

25. Håpnes and Sørensen, "Competition and Collaboration," 189.

26. Peter H. Salus, *A Quarter Century of Unix* (Reading, MA: Addison-Wesley, 1994). For a sense of Unix as a culture, see Don Libes and Sandy Ressler, *Life with UNIX: A Guide for Everyone* (Englewood Cliffs, NJ: Prentice Hall, 1989).

27. Ulrike Erb, "Exploring the Excluded: A Feminist Approach to Opening New Perspectives in Computer Science," in *Women, Work, and Computerization: Spinning a Web from Past to Future*, Proceedings of the 6th International IFIP Conference, Bonn, Germany, May 24–27, 1997, ed. A. Frances Grundy et al. (Berlin and New York: Springer Verlag, 1997), 203–4.

28. Ibid., 206.

29. Lucy Suchman, "Supporting Articulation Work: Aspects of a Feminist Practice of Technology Production," in *Women, Work, and Computerization: Breaking Old Boundaries—Building New Forms*, ed. Alison Adam, Judy Emms, Eileen Green, and Jenny Owen (Amsterdam: Elsevier, 1994), 7–21.

30. Peter Naur and Brian Randell, eds., *Software Engineering: Report on a Conference Sponsored by the NATO Science Committee, Garmisch, Germany, 7th to 11th October 1968* (Brussels: NATO Scientific Affairs Division, 1969), 13. Reprinted as *Software Engineering: Concepts and Techniques, Proceedings of the NATO Conferences*, ed. Peter Naur, Brian Randell, and J. N. Buxton (New York: Petrocelli/Charter, 1976).

31. Perhaps the most prominent is Mary Shaw, professor of computer science at Carnegie-Mellon University and former chief scientist at the Department of Defense–sponsored Software Engineering Institute there in the late 1980s. Shaw earned her reputation in the area of data abstraction but most recently has emerged as a strong proponent of an architectural approach to software development; see Mary Shaw and David Garlan, *Software Architecture: Perspectives on an Emerging Discipline* (Upper Saddle River, NJ: Prentice Hall, 1996).

32. Well, almost and in principle; compare Donald MacKenzie, "The Auto-

mation of Proof: A Historical and Sociological Exploration," *Annals of the History of Computing* 17, no. 3 (1995): 7–29.

33. When I made this point in another context in a talk at the Dibner Institute in 1996, Joel Moses, the provost of the Massachusetts Institute of Technology, commented that a recent study of their program in software engineering had come to essentially the same conclusion and they were revising the curriculum so as to direct students to courses outside computer science.

TEN

Medicine, Technology, and Gender in the History of Prenatal Diagnosis

RUTH SCHWARTZ COWAN

I began my research career before I was conscious of being a feminist, so I have no trouble enumerating the ways in which my research has been affected by my politics over the years since 1970. Feminism has also had a profound effect on the subject on which my research currently focuses: the history of prenatal diagnosis. In this chapter I intend to explore both of these topics. This exploration is made complex by the fact that the topics are conceptually but not historically related to each other; I will try to simplify the analysis by recounting and then analyzing several aspects of the history of amniocentesis and chorionic villus sampling.[1]

Amniocentesis and chorionic villus sampling are techniques for learning something about the genetic and physiological character of a fetus before it is born. Amniocentesis involves removing a small sample of the amniotic fluid which surrounds the fetus inside its mother's womb; this is the fluid that is called "water" when a woman is in labor and her "water breaks." Fetal cells float in amniotic fluid (some are sloughed off the fetus's skin; others are excreted when the fetus urinates); the chromosomes in these cells can be examined through a technique called *karyotyping* (making an enlarged photograph of the chromosomes and then cutting and pasting—which can now be done automatically—the chromosomal figures so that the pairs are arrayed in a fixed sequence), and so can the DNA out of which these chromosomes are made. In addition, the amniotic fluid itself can be subjected to a number of biochemical tests to determine whether the fetus is secreting various substances which are indicators that certain genetic or developmental diseases or disabilities may be present. Amniocentesis cannot be performed until there is a sufficient quantity of amniotic fluid with a sufficient quantity of cells floating in it; this usually happens between the fourteenth and the sixteenth week of a pregnancy, which is roughly midway through the second trimester. Karyotyping cannot be performed unless the fetal cells are cultured, which means that diagnosis of certain conditions (called *chromosomal aberrations*) cannot be made for another one to two weeks.

Chorionic villus sampling is somewhat different; it involves taking a small sample of the projections (called *villi*) which start to form on the *chorion* (one of the fetal membranes) just before the chorion develops into a fully formed *placenta,* implanted in the wall of the mother's uterus. Fetal membranes are *fetal;* the cells in them belong to the fetus, not the mother. This means that the cells removed from the chorionic villi can be analyzed in the same way as the fetal cells in amniotic fluid: by karyotyping and by DNA analysis. Chorionic villus sampling does not yield information about chemicals secreted by the fetus, but it can be performed much earlier in a pregnancy (between the ninth and the eleventh week—but *only* during that window of opportunity), just before the end of the first trimester. Amniocentesis was the first technique for prenatal diagnosis to become widely diffused (in the United States this began happening at the end of the 1970s); chorionic villus sampling followed, roughly a decade later.

FIRST EXAMPLE

Today the condition for which fetuses are most frequently diagnosed is Down syndrome, which can result from one of several different chromosomal aberrations that can be observed by karyotyping. One of the first published accounts of successful and accurate prenatal diagnosis (by amniocentesis) for Down syndrome appeared as a case history in the Letters section of *The Lancet* in 1968.[2] It reports the case of a twenty-nine-year-old pregnant woman known to be a carrier of something which was then called a D/G translocation. A *translocation,* as its name implies, occurs when two chromosomes break and exchange segments; in this case a section of one of the chromosomes in the D group (either chromosome 13 or 14 probably) was exchanged with a section of a chromosome in the G group (most likely chromosome 21). Because they are heterozygotes, carriers of translocations are phenotypically normal, but there is a great risk that their children will have Down syndrome. In this particular case the mother was known to be a carrier because she already had a Down child with the D/G translocation. Apparently the mother agreed to have amniocentesis (in 1968 it was considered an experimental procedure; the technique for culturing fetal cells had only been perfected two years earlier, and the risks of the procedure to mother and fetus had not yet been determined). The same translocation was discovered in the fetal cells. The mother requested a "therapeutic abortion." Postmortem examination of the physical features of the fetus confirmed the diagnosis made through karyotyping of the fetal cells.

The *Lancet* letter ends with a short paragraph that begins with a very peculiar sentence. "The patient greatly wants to have children and is willing to become pregnant again, provided that the same diagnostic technique is applied to her future unborn babies." This sentence is peculiar in two ways. First, the patient speaks in the active voice; this is extremely rare in the medical literature, even in the literature of individual case histories. Second, the diagnostic technology and the abortion are referred to, in essence, as *facilitating* future pregnancies.

In making sense of these peculiarities it helps to remember where we were, in 1968, in terms of abortion policy—and geographically where the authors and the patient were located. The physicians all had appointments in the Department of Obstetrics and Gynecology at the Downstate Medical Center—a teaching and research facility—in Brooklyn, New York; it seems a reasonable assumption that the patient lived somewhere in the New York metropolitan region. Most abortions were still illegal in the state of New York in 1968, although the law did make an exception in the interest of the "life and health" of the mother. In order for a physician to perform a *legal* abortion in New York in those days, a panel of physician peers had to concur that the continuation of the pregnancy was a serious threat to the life or the health (some hospital boards were willing to include *mental health* under this rubric) *of the mother*. That is the reason the term "therapeutic abortion" is used earlier in the paper. In 1968 the abortion reform movement was picking up steam in New York and in other states; legalized abortion was then, as it is now, an extremely sensitive political issue.[3] This may be the reason "Prenatal Diagnosis of Down's Syndrome" was published, somewhat inconspicuously, as a letter to the editor of *Lancet,* a British journal, rather than as a case history article in, say, the *American Journal of Obstetrics and Gynecology*. With a letter to *Lancet* the authors established their priority in a prestigious forum while at the same time avoiding local publicity. Within the medical community physicians were divided—as they remain today— about the wisdom of legalizing abortion; this is almost certainly why the authors of the *Lancet* letter allowed the patient to speak in her own voice *about her desire to have more children* and about the sense in which *prenatal diagnosis will facilitate future pregnancies and also minimize the likelihood of future abortions.* When carriers of this genetic defect become pregnant, the chances of having a normal fetus are still higher than the chances of having an afflicted one.

I read this *Lancet* article very early—probably in 1982 or 1983—in my research into the history of prenatal diagnosis and I filed it away in a folder labeled with the lead author's name. Within the next several years three things happened, although I confess that at this remove I cannot remember the order in which the first two happened: I was invited to give a paper at a workshop on the history and sociology of technological systems; I began to read the literature of feminist epistemology, particularly the work of Sandra Harding; I was asked to give a talk on my research to my colleagues in the English department at the State University of New York at Stony Brook.

When the invitation came to prepare a paper on the application of the concept of technological systems, I was already familiar with the concept as it had been initially developed by Thomas Parke Hughes.[4] Hughes (and many other historians of technology who have adopted his approach) looked at technological systems (networks of people, machines, and social institutions) from the perspective of the people who either created them or managed them—usually male, often inventors, engineers, and businessmen. For my workshop paper I decided, however, to analyze the history of a particular technological system with which I was already familiar (the home heating system) from the perspective of consumers.[5] Consumers are, of course, often female; even when they are not female, they are normalized as female in relation to inventors, engineers, and businessmen: the passive agent in technological change rather than the active agent; the dependent, rather than the independent, variable. Possibly while writing the workshop paper, certainly shortly thereafter, I realized that I had adopted a feminist strategy in the paper, what the philosopher Sandra Harding refers to as a *feminist standpoint* (looking at a subject from a female perspective) combined with *strong objectivity* (keeping to the highest evidentiary standards in drawing conclusions from that perspective).[6] And shortly after *that* I realized that my new research—into the history of prenatal diagnosis—could benefit not only from systems thinking but from feminist systems thinking: the consumers of this new technology were, after all, *all* female.

When analyzed as systems, and analyzed from the perspective of the patient (the person who is *always* female in the case of prenatal diagnosis), both amniocentesis and chorionic villus sampling have six elements: (1) there is an intake medical professional—someone to whom the patient has gone for prenatal care; (2) then that same medical professional or a different one—depending on where the machine is located—examines the fetus by ultrasound; (3) then a procedure is performed on the patient's body—usually, but not always, by a different medical professional—in which some fluid (in the case of amniocentesis) or some tissue (in the case of chorionic villus sampling) is removed; (4) then this fluid or tissue is subjected to a series of laboratory tests; (5) then a report is made to the patient, sometimes by yet another medical professional, called a genetics counselor; (6) then, in some cases, the patient may decide to have an abortion, to terminate the pregnancy—which procedure may be performed by yet another medical professional.

I learned two crucial things from this system analysis of my own research subject. First, that several different medical professionals are involved (obstetrician-gynecologists, radiologists, ultrasound technicians, genetics counselors, nurses) in the history of prenatal diagnosis, in addition to the research physicians who had been, until then, the focus of my studies; if I did not learn something about the history of these professions, I came to understand, I would not be able to write accurately about the *diffusion* of prenatal diagnosis (about which more below). Second, the system analysis taught me to pay attention to something that was absent from all the medical review literature on amniocentesis and prenatal diagnosis. (Review articles are the ways in which medical scientists, in a sense, write the histories of their own fields.) From the scientist's point of view and also from the points of view of many clinicians, the technological system that is prenatal diagnosis ends when the diagnosis is made. But from the *patient's* point of view, the system ends either when good news arrives (meaning that the abortion decision does not have to be made) or when the pregnancy is terminated because bad news has arrived and the mother has decided to have an abortion.

Thus when I took that *Lancet* letter out of its file in order to prepare a talk for my colleagues in the English department, I suddenly realized that its last paragraph was speaking to me—as it must have been speaking to the physicians who had read it in 1968—as much about the history of abortion as it was about the history of prenatal diagnosis. This was the juncture at which I became convinced that these two histories—of prenatal diagnosis and of abortion—were intertwined, *must have been intertwined,* in historically significant ways, ways that I would have to explore if my history of prenatal diagnosis were to be accurate.

In short, my feminism influenced my research, not only by suggesting the topic but by helping me to better understand its complexities (when I realized that I would need to know about the history of several medical professions) and its determinants (when I realized that changes in abortion policy helped explain the pattern in which prenatal diagnosis developed). Sandra Harding is, I believe, right. When we are conscious of our political commitments, we can use them to make our research *better:* more accurate, more insightful, more complex, closer to historical reality, and, for all of those reasons, therefore *more* (rather than less) objective.

SECOND EXAMPLE

A decade before it became possible to diagnose Down syndrome by karyotyping fetal cells in amniotic fluid, it was already possible to diagnose sex through amniocentesis. This diagnosis depended on observation, not of the chromosomes, but rather of a nuclear inclusion called *sex chromatin.* In 1954 a team of Canadian histologists had announced that sex chromatin (which they had discovered, initially, in cat nerve cells) was also present in humans.[7] Within a few months *four* separate research groups, one each in New York, Jerusalem, Minneapolis, and Copenhagen, had announced that they had been able to determine the sex of a fetus before it was born by looking at the fetal cells in the amniotic fluid collected in the delivery room.[8]

There were not many medical geneticists in the mid-1950s, but this piece of news was of enormous significance to the few there were. The only diagnostic technique available to medical geneticists in 1955 was a family history—and it was a profoundly inadequate technique. The only advice a medical geneticist could give was probabilistic: "*If* you decide to have children, *then* there will be a 50-50 (or 25 percent or whatever) *chance* that each baby you have will be afflicted with this or that disease. Do what you think best." This was frustrating advice, for patients and physicians alike.

Prenatal diagnosis for sex, however, held out the possibility not only of a more certain diagnosis but also of an effective therapy for one group of patients, a group that was small in terms of the general population but very large in terms of a medical geneticist's practice: patients with a family history of hemophilia. Hemophilia was known to be a sex-linked disease; in general, women are carriers and men are afflicted. (Since the hemophilia gene is recessive and on the X chromosome, a girl can be born with hemophilia only if her mother is a carrier and her father is a hemophiliac.) If carrier mothers could be offered the possibility of terminating the pregnancies of male fetuses, medical geneticists would have a powerful new tool at their command.

Since the amniotic tap itself (the part of the procedure in which a sample of amniotic fluid is removed from a pregnant woman's uterus) was a relatively well-known and risk-free technique (obstetricians were using it both to ease the difficulties of hydroamnios and to test for Rh sensitivity), several medical geneticists began offering prenatal diagnosis for sex to their patients with a family history of hemophilia. However—and this is the salient point—only one of these physicians had the courage to publish an account of what he was doing: Fritz Fuchs, who was then practicing in Copenhagen.[9] Why?

In the late 1950s Denmark was one of the few countries in the world (the others were Sweden and Norway) in which abortions, especially for what were then called *eugenic* reasons, were legal. Anecdotal and interview evidence that I have been able to collect has led me to the conclusion that Fuchs was not the only medical geneticist who was using amniocentesis and sex chromatin analysis in the pregnancies of women who were carriers of hemophilia. But he was the only person who was able to proceed with the resultant abortions within the letter of the law; hence he was the only person willing to publish his case histories in the 1950s.

Two points are relevant here. The first has already been made: the history of prenatal diagnosis is inextricably connected to the history of abortion. The second pertains to feminist standpoint theory: not all hypotheses generated from a feminist standpoint will turn out to be true. Several radical (or, as they are sometimes called, reproductive) feminists have argued that prenatal diagnosis discriminates against women because it is frequently used, in India and China especially, first to diagnose sex and second to abort female fetuses—and also because people could potentially use it to plan the sex birth order of their children (son first, daughter second).[10] Gena Corea asserts that this form of discrimination could lead to "gynicide"—the elimination of many women and the subordination of the rest. "If many women in the Third World," she writes, "are eliminated through sex predetermination, if fewer firstborn females exist throughout the world . . . then it is indeed gynicide we are discussing."[11]

My research suggests that this hypothesis is wrong, on several counts. First, the intention of the scientists who developed the first successful technique for sex preselection (as we have just seen) was to identify *male* fetuses who were at risk of being hemophiliacs. Second, their intention was to give women the ability to *choose* abortion, thereby expanding the reproductive options of adult women; expansion of options is hard to construe as a form of discrimination or oppression. Third, in many places (Sweden, for example, and several

Indian states), legislatures have voted to prevent medical geneticists from disclosing the sex of fetuses (except in cases of X-linked disease). Fourth, those places where prenatal diagnosis is used to facilitate the abortion of female fetuses are also places in which female infanticide is common. Fifth, there is no evidence that large numbers of people in the developed world have used prenatal diagnosis to design the birth order of their children by sex; neither the sex ratio nor the proportion of firstborns who are female has altered in the decades since prenatal diagnosis became widely diffused. In short, gynicide does not seem to be an immediate or a long-range threat in the developed world and in the developing world, and if it were to occur, prenatal diagnosis could hardly be blamed for it, given how much less expensive it is to expose a female infant than to abort a female fetus. Neither amniocentesis nor chorionic villus sampling was developed with the intention of discriminating or oppressing women, and the diffusion of both techniques has probably lessened, rather than increased, the level of women's subordination. In short, not all hypotheses generated by feminist theory end up being validated by feminist scholarship. Feminists are not univocal; neither is feminist scholarship. For all of these reasons, feminist scholarship is exciting and a joy to pursue.

THIRD EXAMPLE

Diffusion is more important—and more problematic—to historians of technology than to historians of science. With regard at least to twentieth-century science, we know pretty much who the agents of diffusion are: scientists, popularizers of science, funders of science. We also know pretty much what the media are through which scientific findings get diffused: refereed journals, scientific meetings, science journalism. With regard to twentieth-century technology, however, there are many more possible agents and media, depending, of course, on who the consumers of the technology are likely to be; for production technologies the consumers are likely to be business people and engineers; for agricultural technologies the consumers are farmers and agribusiness managers; for domestic technologies the consumers are home owners and gift givers. Each type of consumer is addressed by different agents of diffusion and different media: sales people and trade journals, advertising agencies and women's magazines, to suggest just a few of the many possibilities. There is a large body of information and theory about diffusion, both in sociology and in the history of technology.[12]

Diffusion theory suggests that certain individuals, by acting as leaders, can accelerate the diffusion process; governments are also implicated in diffusion, since their policies can affect the rate and character of diffusion. Advertising agencies and market research companies are also—or can be—agents of diffusion.

In the case of amniocentesis in the United States, two or three women, and their spouses, were—without necessarily intending to be—crucial agents of diffusion. By the early 1970s the women's health movement had become an important aspect of American feminism. Several matters galvanized this movement: abortion reform was certainly crucial, but so was sterilization without consent, childbirth anesthesia, bottle-feeding (as opposed to breast-feeding) of infants, high rates of hysterectomy, and the side effects of oral contraceptives. Women's health activists opposed the medicalization of childbirth in particular and of women's health in general, arguing that the medical profession was fundamentally patriarchal and profoundly misogynist.

The women's health movement coincided with, and participated in, the movement to increase patient autonomy.[13] Stemming from revelations about various medical investigations that had either withheld treatment from some subjects (as in the infamous Tuskegee syphilis study) or caused irreparable physical or emotional damage to other subjects, the movement to increase patient autonomy led to the creation of several quasi-governmental bioethics panels, which themselves led to the creation of both institutional review boards and consent forms.

One other product of both the women's health movement and the movement to increase patient autonomy was a rise in the number of malpractice suits—a technique advocated by attorneys and activists since it empowered patients and allowed redress not just for individual patients but for whole classes of patients. Obstetricians were frequently the target of such suits.

In the United States the events which, possibly more than any others, were responsible for the rapid diffusion of amniocentesis were the settlement in 1978 and 1979 of several lawsuits in which the parents of children born with a disability successfully sued for malpractice when an obstetrician had failed to refer a patient over the age of thirty-five for amniocentesis: the first of the so-called "wrongful life" suits.[14] In one of the first of those cases Dolores Becker, who was thirty-seven years old when she became pregnant, was awarded her child's medical costs for life.[15] The American College of Obstetricians and Gynecologists and the American Academy of Pediatrics subsequently advised their members that, henceforth, they had better offer prenatal diagnostic services or referrals or risk the same kind of suit.[16]

In short, by suing their doctors, several American women had hastened the diffusion of amniocentesis, had acted as agents of diffusion. Ironically, these women were guided by feminist principles of empowerment to take actions that at least some feminists (radical or reproductive feminists) would find reprehensible. All forms of prenatal diagnosis increase the medicalization of pregnancy. Ironically, the rise of the women's health movement—a social movement passionately opposed to the medicalization of pregnancy—is at least partially responsible for the fact that prenatal diagnosis came to be considered a standard aspect of prenatal care, at least in the United States.

CONCLUSION: MULTIPLE IRONIES IN THE INTERSECTION OF GENDER, FEMINISM, AND PRENATAL DIAGNOSIS

Feminists are not univocal on many subjects—and reproductive politics makes for very strange bedfellows, even among feminists. Nowhere are the paradoxes more obvious than in the freighted question of prenatal diagnosis for sex. As amniocentesis and chorionic villus sampling became standard aspects of obstetric practice, patients *without* a family history of a sex-linked genetic disease began inquiring about the sex of their fetuses—and requesting abortions of female fetuses. Many American physicians were appalled and refused to comply with their patients' requests.[17]

Some of these physicians understood that they were caught in what might be called a *feminist paradox*. Physicians who refused to terminate pregnancies for sex selection were compromising the autonomy of their female patients (who were, after all, the people requesting the abortions) *and* placing restrictions on the right to abortion that had been guaranteed in *Roe v. Wade*. Which feminist principle ought to take precedence? Respecting the autonomy of female patients? Or preventing discrimination against female fetuses?

I close with that paradox. As scholars, feminism has given us many gifts. But as actors in history, as consumers and diffusers of twentieth-century reproductive technologies, feminism has also helped to present us with many choices, which are not all that easy either to make or to justify.

NOTES

1. An excellent ethnography and feminist analysis of the current practices of prenatal diagnosis is offered by Rayna Rapp (*Testing Women, Testing the Fetus:*

The Social Impact of Amniocentesis in America [New York: Routledge, 1999]). The focus in this chapter will be on its historical development.

2. Carlo Valenti, Edward J. Schuttle, and Tehila Kehaty, "Prenatal Diagnosis of Down's Syndrome," *Lancet,* July 27, 1968, 220.

3. Kristin Luker, *Abortion and the Politics of Motherhood* (Berkeley and Los Angeles: University of California Press, 1984).

4. Thomas Parke Hughes, "The Electrification of America: The System Builder," *Technology and Culture* 17 (1976): 646–49; Thomas Parke Hughes, *Networks of Power: Electrification in Western Society, 1880–1930* (Baltimore, MD: Johns Hopkins University Press, 1983).

5. The paper that resulted was published in the volume of workshop papers: Ruth Schwartz Cowan, "The Consumption Junction: A Proposal for Research Strategies in the Sociology of Technology," in *The Social Construction of Technological Systems: New Directions in the Sociology and History of Technology,* ed. Wiebe E. Bijker, Thomas P. Hughes, and Trevor J. Pinch (Cambridge, MA: MIT Press, 1987), 261–80.

6. Sandra Harding's views can be sampled in *The Science Question in Feminism* (Ithaca, NY: Cornell University Press, 1986); in her introduction to *Feminism and Methodology: Social Science Issues,* ed. Sandra Harding (Bloomington: Indiana University Press, 1987); and in *Whose Science? Whose Knowledge? Thinking from Women's Lives* (Ithaca, NY: Cornell University Press, 1991).

7. K. L. Moore and M. L. Barr, "Nuclear Morphology, according to Sex, in Human Tissues," *Acta Anatomica* 21 (1954): 197–208.

8. D. M. Serr, L. Sachs, and M. Danon, "Diagnosis of Sex before Birth Using Cells from the Amniotic Fluid," *Bulletin of the Research Council of Israel* 5B (1955): 137; L. B. Shettles, "Nuclear Morphology of Cells in Human Amniotic Fluid in Relation to Sex of Infant," *American Journal of Obstetrics and Gynecology* 71 (1956): 834; F. Fuchs and P. Riis, "Antenatal Sex Determination," *Nature* 177 (1956): 330; E. L. Makowski, K. A. Prem, and I. H. Kaiser, "Detection of Sex of Fetuses by the Incidence of Sex Chromatin Body in Nuclei of Cells in Amniotic Fluid," *Science* 123 (1956): 542.

9. P. Riis and F. Fuchs, "Antenatal Determination of Foetal Sex in Prevention of Hereditary Diseases," *Lancet* 2 (1960): 180.

10. See, for example, Patricia Spallone, *Beyond Conception: The New Politics of Reproduction* (Granby, MA: Bergin and Garvey, 1989); and Gena Corea, *The Mother Machine: Reproductive Technologies from Artificial Insemination to Artificial Wombs* (New York: Harper and Row, 1985).

11. Corea, *Mother Machine,* 206.

12. For one summary, see Everett M. Rogers, *Diffusion of Innovations* (New York: Free Press of Glencoe, 1962).

13. On this subject see David J. Rothman, *Strangers at the Bedside: A History of How Law and Bioethics Transformed Medical Decision Making* (New York: Basic Books, 1991), 142–44.

14. T. D. Rogers, "Wrongful Life and Wrongful Birth: Medical Malpractice in Genetic Counseling and Prenatal Testing," *Science Law Review* 33 (1982): 713.

15. *Becker v. Schwartz* (1978) 46 NY 2d 401, 386 NE 2d 807.

16. Alfred W. Brann and Robert C. Cefalo, eds., *Guidelines for Perinatal Care* (Evanston, IL: American Academy of Pediatrics and American College of Obstetricians and Gynecologists, 1983).

17. See T. M. Powledge and J. Fletcher, "Guidelines for the Ethical, Social, and Legal Issues in Prenatal Diagnosis: A Report from the Genetics Research Group of the Hastings Center, Institute of Society, Ethics, and the Life Sciences," *New England Journal of Medicine* 300 (1979): 168–72; see also John C. Fletcher, "Sounding Board: Ethics and Amniocentesis for Fetal Sex Identification," *New England Journal of Medicine* 301 (1979): 551.

Medicine

ELEVEN

On Bodies, Technologies, and Feminisms

NELLY OUDSHOORN

This chapter is a collection of stories about the complex relationships among bodies, technologies, and feminisms. I will tell three different stories: first, *the Cinderella story,* which tells us how feminists have treated bodies and technologies as neglected stepsisters; second, *the Madonna story,* which describes how female bodies and medical technologies became celebrated and deconstructed in feminist discourse; and third, *the Emperor story,* to show how male bodies gradually came to be included in feminist research. All three stories have a similar plot. I have chosen these stories to illustrate how feminists, myself included, have tried to cope with essentialist views of bodies and technologies.[1]

THE CINDERELLA STORY

Once upon a time feminism was on uneasy terms with everything that smacked of nature, biology, and bodies. In the 1970s, most feminist biologists, like myself, strongly rejected biological explanations of human, and particularly female, nature. We chose this position to oppose antifeminists who suggested that social inequality between women and men is primarily rooted in biological sex differences. "Biology is destiny," and feminists simply have to accept this reality, so they argued.

At this juncture in history, we could have chosen to question this "reality," particularly the assumption that there exists such a thing as a natural body. We might have questioned the status of the biomedical sciences as providers of objective truth and their assumed role of objective arbiters in social debates. Whatever else may have happened in these exciting years, feminists did not take up this challenge. The biomedical sciences were not included in the feminist research agenda.

Feminists did not claim the body for feminist research.[2] Instead we focused our attention on the social sciences. Simone de Beauvoir's argument that "women are made, not born," functioned as a paradigm for feminist scholars in sociology, anthropology, and psychology seeking to analyze the social and cultural contexts of sexual inequality.[3] In the 1970s feminists introduced the concept of gender—as distinct from sex—into the discourse on women. In *Sex, Gender and Society,* the British sociologist Ann Oakley emphasized the relevance of making a distinction between "biological, innate sex differences" and "gender attributes that are acquired by socialization."[4] In this approach, the use of the concept of sex became restricted to biological sex, implicitly or explicitly specified in terms of anatomical, hormonal, or chromosomal criteria. Gender is used to refer to all other "socially constructed characteristics" attributed to women and men, such as specific psychological and behavioral characteristics, social roles, and particular types of jobs.

What actually happened was that feminists, by introducing the sex-gender distinction, reproduced the traditional task division between the social sciences and the biomedical sciences. Feminists assigned the study of sex to the domain of the biomedical sciences and defined the study of gender as the exclusive domain of the social sciences.[5] My point here is not to deny the productivity of the introduction of the sex-gender distinction. The 1970s witnessed the publication of numerous gender studies that revealed the social, cultural, and psychological conditions under which girls and women acquire a feminine role and identity. My argument is that the sex-gender distinction did not challenge the essentialist notion of a natural body. Although the concept of gender was developed to contest the naturalization of femininity, the opposite has happened. Feminist theories of socialization did not question the biological sex of those subjects who become socialized as women; they took sex and the body for granted as unchanging biological realities that needed no further explanation.[6] In these studies, the concept of the sexed body maintained its status as an ahistorical, unproblematic base upon which gender is inscribed. Consequently, the body remained excluded from feminist analysis.

It was in the late 1970s and early 1980s that the body made its first appearance in feminist writings. Historians proposed to include the female body in feminist research. In "La storia della donna," Gianna Pomata challenged the assumption that the female body has a universal, transhistorical essence.[7] Following Pomata, women's history now focused on the particular historicity of women's experiences with their bodies. These studies showed most powerfully how our perceptions of the female body are always historically contingent. "[W]e cannot speak of the feminine body as if it were an invariant presence through history. There is no fixed, experiential base which provides continuity across the centuries."[8]

In addition to historians, anthropologists also came under the spell of the body, bringing an awareness that perceptions of one's body are bound by culture and stressing the cross-cultural diversity of bodily experiences. Each culture attributes different meanings to the female body. Emily Martin's book *The Woman in the Body* extended the anthropological approach to the experiences of women in contemporary American culture. Martin showed how women's imaging of their bodies can vary even within one culture, due to differences in social and economic backgrounds.[9]

Anthropologists and historians provided very powerful insights that challenged the notion of a natural body; however, they only went halfway. These studies focused on experiences with the body and on how these experiences are molded by time and culture. This still left room for the argument that despite differences in bodily experiences, these experiences do refer to a universal, physiological reality, a nonhistorical biological matter.[10] In the experiential approach the facticity and self-evidence of "biological facts" about the body remained unchallenged.[11]

Feminist biologists and historians of science did not hesitate to make this crucial move in exposing the myth of the natural body. Ruth Bleier, Ruth Hubbard, Evelyn Fox Keller, and Helen Longino suggested that anatomical, endocrinological, or immunological "facts" are anything but self-evident.[12] From these feminist scholars I adopted the intellectually challenging and politically relevant notion that no unmediated natural truth of the body can be said to exist. The body is always a signified body. Our perceptions and interpretations of the body are mediated through language, and in our society, the biomedical sciences function as a major provider of this language. This view of the body is linked to a critical reappraisal of the status of biomedical knowledge. If understanding the body is mediated by language, scientists are bound by language as well. Consequently, the assumption that the biomedical sciences are the providers of objective knowledge about the "true nature" of the body could be rejected. This altered my view of science and the world profoundly. What is science all about, if scientists are not discovering reality? In seeking to answer this question, I was inspired by the literature of the emerging field of social studies of science that introduced the powerful idea that scientific facts are not objectively given but collectively created.[13] This provided a radically different perspective on what scientists are doing: scientists are actively constructing reality rather than discovering reality. For the debate about the body, this means that the naturalistic reality of the body as such does not exist, except insofar as it is created by scientists as an object of scientific investigation.[14] The social constructivist approach opened up a whole new line of feminist research, exposing the multiple ways in which the biomedical sciences as discursive technologies (re)construct and reflect our understanding of gender and the body.[15] The body, in all its complexities, thus achieved an important position on the feminist research agenda.

THE MADONNA STORY

The Cinderella story tells us how feminists gradually embraced constructivist approaches to gender, bodies, and technologies in order to contest essentialism. The constructivist approach enabled feminists to go beyond the notion of the natural body. So, exit Cinderella, enter Madonna: in the late 1980s female bodies and medical technologies came to be included in the feminist research agenda. Like Madonna, the Madonna story has many different faces that reflect different feminist strategies to contest essentialism.

The first strategy consists of showing the contingencies in meanings of sex and bodies in medical discourse through the centuries. Feminists who have adopted this strategy have used the myriad ways in which scientists have understood male and female bodies as countermoves to the argument that sex is an unequivocal, ahistorical attribute of the body that, once unveiled by science, is valid everywhere and within every context. Even within the world of science, the domain that claims Objectivity and Universality, bodies appeared to have histories and cultures. This strategy has enriched the literature with beautiful stories. One of these stories is about the construction of sexual Sameness and Difference in medical discourse.

For our postmodern minds it is hard to imagine that for two thousand years male and female bodies were not conceptualized in terms of differences. Medical texts from the ancient Greeks to the late eighteenth century described male and female bodies as fundamentally similar. Women had even the same genitals as men, with one difference: theirs are inside the body and not outside it. In this approach, characterized by Thomas Laqueur as the "one-sex model," the female body was understood as a "male turned inside herself," not as a different sex but as a lesser version of the male body.[16] It was only in the eighteenth century that biomedical discourse began to conceptualize the female body as essentially different from the male body. The long-established tradition that emphasized bodily similarities began to be heavily criticized. In the mid-eighteenth century, anatomists increasingly focused on bodily differences between the sexes and argued that sex was not restricted to the reproductive organs. In nineteenth-century cellular physiology, the medical gaze shifted from the bones to the cells. By the late nineteenth century, medical scientists had extended this sexualization to every imaginable part of the body: bones, blood vessels, cells, hair, and brains.[17] Only the eye seems to have had no sex.[18] Biomedical discourse thus shows a clear shift in focus from similarities to differences, as the female and the male body became conceptualized in terms of opposite bodies with "incommensurably different organs, functions, and feelings."[19]

Following this shift, the female body became the medical object par excellence,[20] emphasizing woman's unique sexual character. Medical scientists now started to identify the "essential features that belong to her, that serve to distinguish her, that make her what she is."[21] The medical literature of this period shows a radical naturalization of femininity in which scientists reduced woman to one specific organ. In the eighteenth and nineteenth centuries scientists set out to localize the "essence" of femininity in different places in the body. Until the mid-nineteenth century, scientists considered the uterus to be the seat of femininity. This conceptualization is reflected in Goethe's statement "Der Hauptpunkt der ganzen weiblichen Existenz ist die Gebaer-mutter."[22]

In the middle of the nineteenth century, medical attention began to shift from the uterus to the ovaries, which came to be regarded as largely autonomous control centers of reproduction in the female animal, while in humans they were thought to be the "essence" of femininity itself.[23] Early in the twentieth century, the "essence" of femininity came to be located not in an organ but in chemical substances: sex hormones. The new field of sex endocrinology introduced the concept of "female" and "male" sex hormones as chemical messengers of femininity and masculinity. This hormonally constructed concept of the body has developed into one of the dominant modes of thinking about male and particularly female bodies. End of this story.

These types of feminist studies illustrate how different bodies have come to life within medical discourse. In these stories, medical science and technology are exposed as powerful agents in the production of the socially crucial categories of male and female bodies. This strategy of historicization is powerful because it disrupts the received wisdom within discourses of sexual Difference and Sameness.

The second feminist strategy to contest essentialist thinking on bodies and technologies is to show how technologies literally transform bodies. Most feminists who have adopted this strategy have studied in vitro fertilization (IVF) technologies and show how IVF technologies have drastically changed female body boundaries. In IVF, fertilization is displaced from the female body and turned into a laboratory event, thus dissolving the distinction between inside and outside the body. Moreover, the use of IVF technologies for male infertility treatment has resulted in the construction of a new object of treatment: the reproductive couple. This new type of patient is problematic, as feminists have argued, because it coincides with the disappearance of women as whole persons from the discourse, a process which threatens the individuality of women. Technoscience discourses on reproduction erase and exclude female bodies and agency.[24] And, finally, IVF has transformed the female body most drastically by shifting the boundaries of the fertility of female bodies: with assisted fertilization techniques women can become pregnant beyond the age of fifty. So, within two decades, body boundaries that were perceived as "natural" for ages have been transgressed and transformed into objects that can be manipulated with an ever growing number of tools and techniques.

Feminists who have adopted this strategy most powerfully show the constructed and increasingly technologically mediated nature of bodies. Medicine has transformed human bodies into cyborgs, to use Donna Haraway's concept: cybernetic organisms resisting the very definition of the natural body.[25] These transformations are not innocent or neutral; on the contrary, in technoscience, bodies, gender identities, and subjectivities are transformed in ways that redefine human agency and power relations between men and women, between doctors and patients, and between humans and things, sometimes for better, sometimes for worse.[26]

The third strategy in contesting essentialist views on technologies and bodies is to show how the naturalistic reality of bodies is created by scientists rather than being rooted in nature. If we reject nature as an external referent, it is important to understand how scientists succeed in convincing us that there exists such a thing as a natural body. The constructivist turn in science and technology studies enabled feminists to go beyond the view that what we perceive as natural is simply dictated by nature and to explore what exactly is required to transform a scientific concept into a natural phenomenon. My research on the development and introduction of the hormonally constructed body concept nicely illustrates this strategy. In *Beyond the Natural Body* I have described the processes of naturalization of the body in terms of the notion of networks of knowledge.[27] From this perspective, knowledge "never extends beyond and outside practices. It is always precisely as local or universal as the network in which it exists. The boundaries of the network of practices define, so to say, the boundaries of the universality of medical knowledge."[28] In this view, the successes and failures in scientists' quest for universal knowledge are ascribed to the extent to which they are successful in creating networks. In my research I have used this constructivist network approach to analyze how the female body, rather than the male body, has become increasingly subjected to hormonal treatment. Why do we talk about female hormones, female pills, and female bodies and not about male hormones and male bodies? Or to rephrase this question in constructivist terms: why and how are claims about female bodies and hormones made more persuasive and attract more support than claims about male bodies and hormones? My book describes how the networks that evolved around statements about female sex hormones were much more extensive and substantial than the networks around male sex hormones. In the case of female sex hormones, laboratory scientists and pharmaceutical companies did not have to start from scratch. They could rely on an already organized medical practice that could easily be transformed into an organized market for their products. The gynecological clinic functioned as a powerful institutional context that provided an available and established clientele with a broad range of diseases that could be treated with hormones.

Knowledge claims about male sex hormones were more difficult to link with relevant groups outside the laboratory. The production as well as the marketing of male sex hormones was constrained by the lack of an institutional context comparable to the gynecological clinic: men's clinics specializing in the study of the male reproductive system did not exist in the 1920s. Although there existed a potential audience for the promotion of male sex hormones, this audience was not embedded in any organized market or resource network. In my book I conclude that it was this asymmetry in organizational structure that made the female body into the central focus of the hormonal enterprise. Sex endocrinologists depended on these organizational structures to provide them with the necessary tools and materials. These differences in institutionalization between the female and the male body constituted a crucial factor in shaping the extent to which knowledge claims could be transformed into universal facts. The institutionalization of practices in a medical specialty transformed the female body into an easily accessible supplier of research materials, a convenient guinea pig for tests, and an organized audience for the products of science. These established practices facilitated a situation in which the hormonally constructed female body concept acquired its status as a universal, natural phenomenon. My research on sex hormones thus shows how we can go beyond the naturalistic reality of bodies by exposing the concrete, often very mundane, human activities that go into discourse building.

THE EMPEROR STORY

The Madonna story shows that feminists have been very creative and productive in contesting essentialism in medical discourse. There is, however, a serious problem, not only in the Madonna story but also in the Cinderella story. Both stories focus almost exclusively on female bodies. If we believe these feminist stories, it seems as if only female bodies have been subjected to historical and cultural shifts in meanings and practices in medical discourse and culture at large. The male body appears as a stable category, untouched by time and place. In feminist discourse the male body maintains its naturalness: it is not a construct; it simply exists.

By focusing too exclusively on female bodies, feminists unwittingly reproduce the tradition in medical discourse which presents female bodies as exotic, as the Other, as bodies that need to be scrutinized and explained to exist.[29] Feminist discourses thus reinforce the dominant image of men as the unmarked sex: male bodies and masculinities do not need to be questioned. There are, of course, good reasons for focusing on female bodies. A critical deconstruction of medical discourses on female bodies is a very important strategy for feminists concerned with monitoring and criticizing the problematic consequences of the medicalization of female bodies. The problem, however, is that by doing so we still grant the Emperor his clothes, in this case the "naturalness" that protects him against critical and deconstructive stories. I think it is time to adopt a new strategy in feminist studies of medical technologies. Let us turn the feminist gaze to the male body and begin to undress the Emperor.[30] The example of male contraceptive technologies shows the usefulness of the Emperor strategy.

Medical discourses on male contraceptive technologies are an important arena for feminist analysis because they show vividly how essentialist representation strategies are used to reproduce and strengthen hegemonic cultural stereotypes of masculinities and male bodies.[31] In the scientific and journalistic literature on male contraceptive technologies, the authors adopt four different representation strategies. First, male bodies are represented as having a reproductive system that is much more complex than that of women. In women, so the argument goes, contraception requires interference with ovulation, which occurs "only once each month." In men, on the other hand, contraception requires intervention in the production of "tens of millions of spermatozoa each day throughout adult life."[32] This representation strategy is used in many texts in scientific journals and newspapers as a tool to convince the reader that the "delay" in the development of male contraceptives is caused by the complex nature of male bodies and not by male bias (as has been claimed by feminist health advocates).

Second, male bodies are represented as being oversensitive to pain. This representation of male bodies is particularly dominant in journalistic accounts of the development of new male contraceptives, as exemplified in the media coverage of the World Health Organization's (WHO) press release on the clinical testing of a new male contraceptive injection.[33] Although the leading actors in the new field of male contraceptive technologies, most notably the WHO, try to use the news media to articulate the acceptability of new male contraceptive methods, most journalists adopt a highly critical attitude toward the new technology. Whereas the WHO presents its newly developed hormonal contraceptive injection as a very promising, highly effective new male contraceptive, the news media (at least in the Netherlands) tell stories in which side effects and pain are the central narrative elements. A leading Dutch journal surveyed the opinions of its readers about the new male contraceptive on trial, and one female respondent concluded: "They need a weekly injection in the buttock? Oh, forget it. There is no man who will do that. Men are oversensitive to injections." And a male physician in the same journal stated: "An injection is painful and men are ten times more sensitive to pain than women, I have been told. So I don't see a breakthrough for a male contraceptive."[34] Dutch newspapers thus construct the image of a painful technology and of oversensitive men. This representation strategy is not innocent; on the contrary, journalists use this strategy to convince their readers that the majority of Dutch men will never use the new male contraceptive.

A third dominant representation strategy in media discourse about male contraceptive technologies is the construction of male identities in terms of unreliability. Under the headline "Where are the cheers for the buttock-injection?" one Dutch journalist told his readers that the male contraceptive injection may be just as reliable as the female pill but simultaneously raised the question "How reliable are men?" This journalist cites the results of a survey among the female visitors to a

Dutch family-planning clinic to convince his readers that women will never accept this technology because they think that men are not responsible enough to prevent unplanned pregnancy.[35] Again, this representation strategy is used to denounce the development of new male contraceptives. Finally, scientists and journalists represent male bodies as vulnerable bodies on which no tinkering is allowed. These representations are implicitly present in the regulations for the clinical testing of contraceptives and in discussions on their side effects. In recent years, several lines of research have been abandoned due to indications of slight adverse effects of contraceptive compound.[36]

This short impression of representation strategies in male contraceptive discourse shows how essentialist views are used to argue that nature dictates the type of technologies we can make. We have thirteen new female contraceptive methods, all developed after World War II, and no new male contraceptives in the course of four hundred years, because of the perception of a different nature in male and female bodies. I want to tell another story about the "failures and delays" in the development of new male contraceptives and the "unreliable nature" of male bodies and identities. In my research I want to go beyond these essentialist views of male bodies and masculinities by exposing how these notions are the results rather than the causes of medical discourses on male bodies. I want to show how male bodies and masculinities gain their specific meanings only in locally and historically specific processes.

WHERE ARE THE POLITICS?

Reflecting on the three stories I have told, we can conclude that feminists have been rather effective in undermining the assumption that bodies are natural, knowable, and essential. But what about the politics of feminist studies of medical technologies? In the last few years there has been a debate among feminists and constructivists on the incompatibility of postessentialism with political commitment. How can we make normative judgments of the benefits or adverse effects of medical technologies if the "truth" no longer exists? Sometimes constructivists and postessentialist feminists are accused of only playing nice academic games. Postessentialist approaches in feminist studies silence feminist criticism and turn feminists and constructivists into "uncommitted and untrustworthy voyeurs at the scene of the crime."[37]

Where are the politics of feminist scholarship in the late 1990s? I think that recent feminist research is not just a depoliticized, academic game. Contesting essentialism in biomedical discourse is still badly needed as an antidote to all these infectious tales of heroic scientists discovering the realities of human nature. We still need critical deconstructions of the processes that shape science, technologies, and bodies to show that science and technology can take many forms. Medical technologies do not necessarily have to be the way they actually are. Even in the new millennium, it is important for feminists and constructivists to keep telling stories about other, silenced or dreamed technologies. Investigating how gender identities, body boundaries, and power relations are construed and figure in medical technologies remains an important feminist project because "medicine is the place where we suffer (or relish) technoscience directly and deeply."[38] This is ultimately a political project because it aims to understand how we can change and alter the power structures in our technoscientific culture.

It must be clear by now that I, as a feminist and a constructivist, strongly reject the suggestion that postessentialist approaches silence feminist criticism; on the contrary, they open up many new opportunities for feminist politics. What has changed through the embracement of constructivism by feminism is the role of the feminist critic. In the 1970s and 1980s feminist criticism consisted of saying yes or no to new medical technologies, or of remaining silent in case of disagreement on the positive or adverse effects of these technologies. By adopting the *politics of contesting technologies,* feminists positioned themselves outside mainstream discursive arenas. I think that constructivism enables feminists to extend their politics to one of *creating meanings* for bodies and technologies. If we adopt the idea that "bodies are heterogeneously constructed by individuals and collectivities situated differently in terms of time and space," we should position ourselves as active participants in the construction of meanings.[39] As Haraway argues, feminism in the 1990s is a "project for the reconstruction of public life and public meanings; feminism is therefore a search for new stories, and so for a language which names a new vision of possibilities and limits."[40]

The role of the feminist critic in the 2000s can thus be formulated in terms of adding new stories to the repertoire of bodily understandings. By doing research on the sociology of male contraceptive technologies, I will add my own story to the many stories that are told about technologies, male bodies, and male identities.[41] My story will in turn become part of the discussion of male contraceptives, "one more interpretation in the field of struggle."[42] One way of writing our stories is to subvert the conventions of discourse by authorizing other voices to speak. Instead of reproducing the hierarchical order of voices by prioritizing what medical scientists have to say about bodies, we can add the voices of feminists, journalists, and fiction writers as different but equally important actors in the construction of discourse about bodies and technologies.[43] My research includes the voices of feminists, journalists, politicians involved in population policies, men who try to develop alternative, low-tech male contraceptive methods outside the academic arena, and so on. I want in this way to expose dissenting voices that refute commonly shared myths and beliefs about male bodies and identities. The excitement of male contraceptive discourse lies in the fact that dissenting voices are not restricted to feminists. Respectable and established organizations such as the WHO, and even biomedical scientists, are active participants in the construction of new meanings for male bodies and identities. So, to paraphrase Bijker: do not despair; there is life after constructivism for feminist politics, even within the mainstream discursive arena.[44]

NOTES

1. Parts of this chapter, particularly parts of the texts in the Cinderella story and the Madonna story, were previously published in Nelly Oudshoorn, *Beyond the Natural Body: An Archaeology of Sex Hormones* (London and New York: Routledge, 1994). A longer version of this chapter was presented at the workshop "New Directions in the Social and Historical Study of Science and Technology," Copenhagen, Dec. 4, 1996. The French version of this chapter has been published in Delphine Gardey and Ilana Löwy, eds., *L'invention du naturel: Les sciences et la fabrication du féminin et du masculin*, Histoire des sciences, des techniques et de la médicine (Paris: Editions des archives contemporaines, 2000), 31–45.

2. This does not mean that feminists completely ignored the biomedical sciences. Feminists involved in the women's health movement criticized the medical profession for not paying enough attention to women's health issues; however, women's health advocates did not question the biological and technological determinism in medical discourse, which is the focus of my chapter.

3. For a more comprehensive analysis of the attitudes of early feminism toward biology, see Lynda Birke, *Women, Feminism and Biology: The Feminist Challenge* (Brighton, Sussex: Wheatsheaf Books, 1986), 1–13.

4. Ann Oakley, *Sex, Gender and Society* (London: Maurice Temple Smith, 1972).

5. For a more detailed analysis of these strategies in feminist research, see Annemarie Mol, "Baarmoeders, pigment en pyramiden," *Tijdschrift voor vrouwenstudies* 9 (1988): 276–90.

6. Ibid.; Barbara Duden, *The Woman beneath the Skin: A Doctor's Patients in Eighteenth Century Germany,* trans. Thomas Dunlap (Cambridge, MA: Harvard University Press, 1991).

7. Gianna Pomata, "La storia della donna: una questione di confine," *Il monde contemporaineo* 10 (1983): 55–107. Dutch translation: Gianna Pomata, "De geschiedenis van vrouwen: een kwestie van grenzen," *Socialistisch-feministische teksten* 10 (1987): 61–113.

8. Mary Jacobus, Evelyn Fox Keller, and Sally Shuttleworth, introduction to *Body/Politics: Women and the Discourses of Science,* ed. Mary Jacobus, Evelyn Fox Keller, and Sally Shuttleworth (New York and London: Routledge, 1990), 4.

9. Emily Martin, *The Woman in the Body: A Cultural Analysis of Reproduction* (Boston, MA: Beacon, 1987).

10. Duden, *The Woman beneath the Skin.*

11. With one major exception: in *The Woman in the Body,* Emily Martin includes a deconstruction of medical views of the reproductive functions of the female body.

12. Ruth Bleier, *Science and Gender: A Critique of Biology and Its Theories on Women* (New York: Pergamon Press, 1984); Ruth Bleier, ed., *Feminist Approaches to Science* (New York: Pergamon Press, 1986); Ruth Hubbard, " 'The Emperor Doesn't Wear Any Clothes': The Impact of Feminism on Biology," in *Men's Studies Modified: The Impact of Feminism on the Academic Disciplines,* ed. Dale Spender (New York: Pergamon Press, 1981), 213–37; Ruth Hubbard, Mary Sue Henifin, and Barbara Fried, eds., *Biological Woman: The Convenient Myth* (Cambridge, MA: Schenkman, 1982); Evelyn Fox Keller, "Feminism and Science," *Signs: Journal of Women in Culture and Society* 7 (1982): 589–95; Evelyn Fox Keller, *Reflections on Gender and Science* (New Haven, CT: Yale University Press, 1985); Helen Longino and Ruth Doel, "Body, Bias, and Behavior: A Comparative Analysis of Reasoning in Two Areas of Biological Science," *Signs: Journal of Women in Culture and Society* 9 (1983): 207–27; Helen Longino, *Science as Social Knowledge: Values and Objectivity in Scientific Inquiry* (Princeton, NJ: Princeton University Press, 1990). In contrast to feminist sociologists, feminist biologists considered the body to be relevant for the feminist research agenda. Actually, feminist biologists have adopted this position from the very beginning of the debate about sex and gender. My account of the history of feminist studies should therefore not be read as a story of continuity and progress. There have been, and still are, many different positions in the debate about sex, gender, and the body.

13. See, among others, Wiebe E. Bijker, Thomas P. Hughes, and Trevor J. Pinch, eds., *The Social Construction of Technological Systems: New Directions in the Sociology and History of Technology* (Cambridge, MA: MIT Press, 1987); Wiebe Bijker and John Law, *Shaping Technology—Building Society: Studies in Sociotechnical Change* (Cambridge, MA: MIT Press, 1992); G. Nigel Gilbert and Michael Mulkay, *Opening Pandora's Box: A Sociological Analysis of Scientists' Discourse* (Cambridge: Cambridge University Press, 1984); Bruno Latour and Steve Woolgar, *Laboratory Life: The Social Construction of Scientific Facts* (Beverly Hills, CA, and London: Sage, 1979); Bruno Latour, *Science in Action: How to Follow Scientists and Engineers through Society* (Cambridge, MA: Harvard University Press, 1987).

14. Duden, *The Woman beneath the Skin.*

15. Susan Bell, "Changing Ideas: The Medicalization of Menopause," *Social Science of Medicine* 24 (1987): 535–42; Birke, *Women, Feminism and Biology;* Bleier, *Science and Gender;* Bleier, *Feminist Approaches to Science;* Adele Clarke, "Women's Health over the Life Cycle," in *Women, Health, and Medicine in America: A Historical Handbook,* ed. Rima D. Apple (New York: Garland Press, 1990), 3–39; Anne Fausto-Sterling, *Myths of Gender: Biological Theories about Women and Men* (New York: Basic Books, 1985); Donna Haraway, "In the Beginning Was the Word: The Genesis of Biological Theory," *Signs: Journal of Women in Culture and Society* 6 (1981): 469–81; Donna Haraway, *Primate Visions: Gender, Race, and Nature in the World of Modern Science* (New York and London: Routledge, 1989); Claudia Honegger, *Die Ordnung der Geschlechter: Die Wissenschaften vom Menschen und das Weib* (Frankfurt and New York: Campus Verlag, 1991); Hubbard, " 'The Emperor Doesn't Wear Any Clothes' "; Janet Sayers, *Biological Politics: Feminist and Anti-feminist Perspectives* (New York: Tavistock Publications, 1982); Londa Schiebinger, "Skeletons in the Closet: The First Illustrations of the Female Skeleton in Nineteenth-Century Anatomy,"

Representations 14 (1986): 42–83; Londa Schiebinger, *The Mind Has No Sex? Women in the Origins of Modern Science* (Cambridge, MA: Harvard University Press, 1989); Marianne van den Wijngaard, "Acceptance of Scientific Theories and Images of Masculinity and Femininity," *Journal of the History of Biology* 24 (1991): 19–49; Marianne van den Wijngaard, "Reinventing the Sexes: Feminism and Biomedical Construction of Femininity and Masculinity, 1959–1985" (Ph.D. diss., University of Amsterdam, 1991).

16. Thomas Laqueur, *Making Sex: Body and Gender from the Greeks to Freud* (Cambridge, MA: Harvard University Press, 1990), 8, 4.

17. Schiebinger, "Skeletons in the Closet."

18. Honegger, *Die Ordnung der Geschlechter.*

19. Laqueur, *Making Sex,* 62.

20. Michel Foucault, *Histoire de la sexualité,* vol. 1, *La volonté de savoir* (Paris: Gallimard, 1976).

21. Laqueur, *Making Sex,* 5.

22. Georg Prochaska, *Physiologie oder Lehre von der Natur des Menschen* (Vienna: Beck, 1820), 64–66, as quoted in Victor Cornelius Medvei, *A History of Endocrinology* (Boston: MTP Press, 1982), 213.

23. Catherine Gallagher and Thomas Laqueur, eds., *The Making of the Modern Body: Sexuality and Society in the Nineteenth Century* (Berkeley and Los Angeles: University of California Press, 1987).

24. Marta Kirejczyk and Irma van der Ploeg, "Pregnant Couples, Medical Technology, and Social Constructions around Fertility and Reproduction," *Issues in Reproductive and Genetic Engineering* 5 (1992): 113–25; Irma van der Ploeg, "Hermaphrodite Patients: In Vitro Fertilization and the Transformation of Male Infertility," *Science, Technology, and Human Values* 20 (1995): 460–82.

25. Donna Haraway, "Manifesto for Cyborgs: Science, Technology, and Social Feminism in the 1980s," *Socialist Review* 80 (1985): 65–108. Reprinted as "A Cyborg Manifesto: Science, Technology, and Socialist-Feminism in the Late Twentieth Century," in *Simians, Cyborgs, and Women,* by Donna Haraway (New York: Routledge, 1991), 149–82.

26. Marc Berg and Monica Casper, "Constructivist Perspectives on Medical Work: Medical Practices and Science and Technology Studies," in *Science, Technology, and Human Values* 20 (1995): 395–408.

27. Oudshoorn, *Beyond the Natural Body.*

28. Bernike Pasveer, "Shadows of Knowledge: Making a Representing Practice in Medicine—X-ray Pictures and Pulmonary Tuberculosis, 1895–1930" (Ph.D. diss., University of Amsterdam, 1992), vii.

29. For a similar conclusion, see Laurence Goldstein, ed., *The Female Body: Figures, Styles, Speculations* (Ann Arbor: University of Michigan Press, 1991), x. This does not imply that male bodies have been completely absent from feminist analysis of medical discourse. Some feminist scholars have included male bodies in their analysis as a strategy to show differences in representations of male and female bodies, for example, in hormonal theories and therapies (Oudshoorn, *Beyond the Natural Body*), in fertility studies (Naomi Pfeffer, "The Hidden Pathology of the Male Reproductive System," in *The Sexual Politics of Reproduction,* ed. Hilary Homans [London: Gower, 1985], 30–44), and in anatomical representations of female and male genitalia (Lisa Jean Moore and Adele Clarke, "Clitoral Conventions and Transgressions: Graphic Representations in Anatomy Texts, c. 1900–1991," *Feminist Studies* 21 [summer 1995]: 255–301). Most attention, however, is given to representations of female bodies.

30. Ruth Hubbard has used the fairy tale "The Emperor's New Clothes" as the title of a paper in which she analyzes the impact of feminism on the biological sciences. Hubbard encouraged feminists to question the "scientifically proven

facts of women's biology." See Hubbard, " 'The Emperor Doesn't Wear Any Clothes.' "

31. Hegemonic stereotypes are those interpretations of masculinities that are perceived as self-evident compared to other expressions of masculinity. See R. W. Connell, *Gender and Power: Society, the Person, and Sexual Politics* (Stanford, CA: Stanford University Press, 1987).

32. Donald W. Fawcett, "Prospects for Fertility Control in the Male," in *Hormonal Contraceptives, Estrogens, and Human Welfare,* ed. Marian Cleeves Diamond and Carol Cleaves Korenbrot (New York: Academic Press, 1978), 57.

33. Press Release, WHO, spring 1996.

34. Both quotations from *De volkskrant,* Apr. 4, 1996.

35. Annemiek Leclaire, "Waar blijft het gejuich om de bilprik?" *HP/De Tijd* 15 (1996): 8–11.

36. For a more detailed analysis of the role of the media in the development of new male contraceptive technologies, see Nelly Oudshoorn, "On Bodies, Technologies, and Pain: The Testing of Male Contraceptives in the Media and the Clinic," *Science, Technology, and Human Values* 24, no. 2 (1999): 265–89. For an analysis of the role of the WHO in contraceptive development, see Nelly Oudshoorn, "The Role of the WHO as an Intermediary Organization in Contraceptive Development," *Social Studies of Science* 27 (1997): 41–72.

37. Rosalind Gill, "Power, Social Transformation, and the New Determinism: A Comment on Grint and Woolgar," *Science, Technology, and Human Values* 21 (1996): 350.

38. Monica Casper and Marc Berg, "Constructivist Perspectives on Medical Work: Medical Practices and Science and Technology Studies," *Science, Technology, and Human Values* 20 (1995): 396.

39. Moore and Clark, "Clitoral Conventions," 257.

40. Haraway, "A Cyborg Manifesto," 175.

41. See Nelly Oudshoorn, *Designing Technology and Masculinity: A Biography of the Male Pill* (New York: Routledge, forthcoming).

42. José van Dijck, *Manufacturing Babies and Public Consent: Debating the New Reproductive Technologies* (London: Macmillan, 1995), 60.

43. Ibid.

44. Wiebe Bijker, "Do Not Despair: There Is Life after Constructivism," *Science, Technology, and Human Values* 18 (1993): 113–29.

TWELVE

Rationality, Feminism, and Mind

EMILY MARTIN

Culture/nature, mind/body, rational/irrational, reason/passion: these dichotomous pairs, whose first term is often given higher value than the second, have lain at the heart of Western reason since the Enlightenment at least, helping to shore up invidious distinctions based on gender, race, sex, or nationality. Feminist analyses have examined historically how gender hierarchies gain strength, for example, when adult males are accorded the realm of culture, mind, rationality, and reason, while women and children are relegated to the realm of nature, body, irrationality, and passion.[1] Feminist analyses have also shown that race hierarchies can be strengthened if white people (or other dominant racialized groups) are granted a "natural" association with culture, mind, rationality, and reason, leaving black people (or other racially contrasted groups) to manage where they fit best, in the realm of nature, body, irrationality, and passion. Toni Morrison describes her experience of reading Marie Cardinal's *The Words to Say It,* an autobiographical account of how this white French girl who had grown up in Algeria collapses into madness. This is what Cardinal wrote:

> My first anxiety attack occurred during a Louis Armstrong concert. I was nineteen or twenty. Armstrong was going to improvise with his trumpet, to build a whole composition in which each note would be important and would contain within itself the essence of the whole. I was not disappointed: the atmosphere warmed up very fast. . . . The sounds of the trumpet sometimes piled up together, fusing a new musical base, a sort of matrix which gave birth to one precise, unique note, tracing a sound whose path was almost painful, so absolutely necessary had

its equilibrium and duration become; it tore at the nerves of those who followed it.

My heart began to accelerate, becoming more important than the music, shaking the bars of my rib cage, compressing my lungs so the air could no longer enter them. Gripped by panic at the idea of dying there in the middle of spasms, stomping feet, and the crowd howling, I ran into the street like someone possessed.

Morrison remembers smiling when reading this passage, in part because Cardinal's recollection of the music had such immediacy and "partly because of what leaped into my mind: what on earth was Louie playing that night? What was there in his music that drove this sensitive young girl hyperventilating into the street," feeling "like someone possessed"? Morrison muses upon the "way black people ignite critical moments of discovery or change or emphasis in literature not written by them" and the "consequences of jazz—its visceral, emotional, and intellectual impact on the listener," which in this case tipped Cardinal from sanity to madness, from rationality to irrationality.[2]

Just as distinctions between rationality and irrationality can play a role in the ranking of people, they can also play a role in the ranking of nation-states. Hugh Gusterson has shown that in literature on economic development, First World countries are depicted as mature, reliable, stable, trustworthy, law-abiding, rational, and logical (in other words, acting like adult men) while Third World countries are depicted as immature, unreliable, volatile, untrustworthy, lawless, irrational, and emotional (in other words, acting like childish women).[3] As for nation-states, so for the disciplines: Avery Gordon lays the groundwork for understanding how in the contemporary social sciences, the producers of rigorous, logical, robust, hard, scientific knowledge about the real world (economists or quantitative sociologists) often attempt to consign to the outer darkness—I use the term advisedly—imposters (cultural anthropologists or sociologists) who pretend to produce knowledge about social worlds while merely telling loose, emotional, delicate, or soft stories.[4]

Feminist efforts to understand how the dichotomies of gender and race are anchored, in order to loosen them, have taken a number of forms. Some, like Elizabeth Grosz, have turned to analysis and validation of the subordinated and excluded term, the body, which is "the unacknowledged condition of the dominant term, *reason.*"[5] Others, like Gusterson or Gordon, argue that those associated with the dominant terms of the dichotomies seek to maintain their separation and distance from the subordinated terms precisely because they unconsciously know and fear the existence of the subordinated qualities within themselves. For Gusterson, the First World "self" in effect contains all the womanly and childish characteristics it imputes, out of fear, to the Third World "other"; for Gordon, the stance that sociology can provide an "unproblematic window onto a more rather than less secure reality" is haunted by a "seething presence," "one form by which something lost, or barely visible, or seemingly not there to our supposedly well-trained eyes, makes itself known or apparent to us."[6] The ghost haunting sociology is that "[t]he real itself and its ethnographic or sociological representations are also fictions, albeit powerful ones that we do not experience as fictional, but as true."[7]

Taking another tack, the one I will follow in this paper, feminist scholars in science studies have argued that the tidy boxes in which the hoary old dichotomies have appeared to be contained for hundreds of years are coming apart.[8] There is increasing awareness, in part through ethnographic research in scientific and nonscientific contexts in which knowledge is produced, that science is less a pure realm of rational, logical knowledge than a porous field transected by many other cultural contexts and forces, which are partners in the production of knowledge.[9] In realms like biomedicine and cyberspace, the walls holding apart the mind and the body, human and machine, rationality and irrationality, are being breached, allowing a whole host of hybrid, cyborg, partial, or amalgamated entities to emerge into the light of day. Creative new possibilities thus open on all sides, pointing the way toward different and perhaps less injurious distinctions. In tune with this flux, there is evidence that change, multiplicity, and flexibility are becoming part of normative conceptions about the nature of personhood in Euramerican cultures. Over the decades since the 1940s, cultural conceptions of the ideal self and body have been becoming less autonomous, isolated, and defended, and more permeable, vulnerable to the outside, and embedded inextricably in complex systems (and in turn within wider systems).[10]

RECUPERATING THE IRRATIONAL

If something as fundamental to the Enlightenment conception of reason as the kind of subject who can know is undergoing dramatic change, the question arises whether this change might extend to recuperation of the irrational, primitive, unconscious, and illogical. Would permeable, vulnerable, and contextually embedded selves not have knowledge that would be figured as irrational or illogical by the canons of Enlightenment reason? Can such "irrational" knowledge gain social validity, and if so, how would it be refigured in the process?

I have been considering this question ethnographically, as part of a larger project that looks at shifts in the value attached to cognitive and emotional states involving lability that occur in both the sciences of the mind and popular culture. Until recently, states involving continuous cognitive or emotional shifting have been understood as illnesses, liabilities, habits of mind that get in the way of forming the kind of self—stable, steady, developing on a unilinear track—most valued in the citizens of modern nation-states. Recently, the value attached to one quintessential category of instability, mania, seems to be shifting, and its status as a liability is perhaps growing less stable. In the sketch below, I will look at ways mania is growing less substantial as a separate and stigmatized category and is instead becoming understood as a cultural construct, a category made in relation to its immediate context and liable to change and shift accordingly. I will also point out ways in which the category mania is simultaneously gaining concreteness, timelessness, and facticity by being attached to particular physical causes that lie behind and explain its presence. Finally, I will consider how the tension between these views ("mania" as a culturally constructed way of being in the world vs. mania as caused by a determinant physical source) bears on what I think may be a kind of "colonization" of certain manic states, which has to do, among other things, with the production of a labile self who seems best "fit" for the environment of late capitalist development.

MANIA SHIFTS GROUND

One way in which the category mania is shifting is that it is coming to be seen as part of a sliding scale of states: even in the official manual of the American Psychiatric Association, the *Diagnostic and Statistical Manual IV (DSM IV)*, there are degrees of being bipolar, including a less severe form called bipolar 2—with hypomania, a less extreme form of mania. Recently another intermediate category has appeared, called "hyperthymia." Peter Kramer, author of *Listening to Prozac*, describes it in a recent *U.S. News and World Report* editorial: "Personality researchers have noted that 'hyperthymic' men—those with a constant upbeat mood, two steps short of mania—have high sexual appetites. Hyperthymia is a personality style thought to have strong biological roots, perhaps related to the biology of manic-depressive illness. Certainly a number of politicians are hyperthymic. They are energetic, optimistic, and decisive, and they require little sleep."[11] Hyperthymia is a temperament, signaled by an individual's typical emotional behavior over an extended period of time.[12] Hyperthymics are the individuals who apply themselves successfully to business, the politicians, those who enjoy being with the crowd and who see themselves as people of action and the leaders of their communities. Thus, although the extreme moods of mania and melancholia may seem separate and apart from common experience, in fact the spectrum of human emotion is broad. "Mood states that have great benefit in some walks of life coexist with others that can do great harm."[13] Peter Whybrow thus suggests that there is an extraordinary and terrifying continuity between the normal self and the madness of manic-depression.[14] This way of speaking of a range of normal and abnormal states as close cousins (a term Whybrow uses) echoes a technique used by the linguist Roman Jakobson, who analyzed the structure of many forms of aphasia. He saw all forms of aphasic alteration in speech, as well as child language, as fully part of language. The differences could be called differences of style. "In manipulating these two [particular] kinds of (linguistic) connection. . . . [an aphasic] exhibits his personal style, his verbal predilections and preferences."[15] Jakobson considered that every aphasia is a style.

Another way the category mania is shifting is that mood disorders involving mania are not only becoming classified as continuous with the normal but undergoing a dramatic change in valence. Accounts of manic-depression have recently been flooding the press, the best-seller list, and the airwaves. Manic-depression has fueled the plot of a series of detective novels (*A Child of Silence* and others by Abigail Padgett), a prison escape novel and movie (*Green River Rising,* 1994), a southern novel (*Sights Unseen,* 1995), plots of the television program *The X Files,* and a memoir on the best-seller list for many weeks (*An Unquiet Mind,* 1995).[16]

What is of compelling interest here is that manic-depression seems to be in the process of redefinition from being a disability to being a strength. These new claims, highly specific to the present social context, can hardly be taken at face value, especially by those who live with manic-depression, but this makes them no less interesting. The psychiatrist who is the author of the memoir I mentioned above, Kay Jamison, takes great pains to describe the positive aspects of manic-depression alongside the negative.[17] Manic-depression entails a "finely wired, exquisitely alert nervous system."[18] These thought processes are characteristic of the manic phase: "fluency, rapidity, and flexibility of thought on the one hand, and the ability to combine ideas or categories

of thought in order to form new and original connections on the other
. . . rapid, fluid, and divergent thought."[19] In a number of books and
Internet sites, there are lists of famous and influential people whose
diaries, letters, and other writings indicate that their manic-depression
played a role in the enormously creative contributions they made to
society. Jamison lists over two hundred composers, artists, and writers
who arguably had some version of manic-depression, from T. S. Eliot
and Edna St. Vincent Millay, to Georgia O'Keeffe, Edvard Munch,
and Jackson Pollock.[20] An important question I can only allude to here
is whether these developments trade on what Michael Bernstein calls
a dangerous tendency toward celebrating excess, or trading on the appeal of the "self declared marginal and powerless."[21]

Manic-depression is said to entail a "distorted" sense of time and
space: "delusions" abound in the manic phase. Objects seem to merge,
to flow into each other. Shapes shift. Any ordinary thing can change
into something else and then something else again. The process happens unbidden and is not necessarily sinister at all. It can be as interesting as watching things morph in a movie! Inside and outside the self
are blurred.[22] Thought is rapid and flighty, jumping from one thing to
another.[23]

As rapidly as they are described, however, these distortions become
assets. In the present environment, in which time and space are in many
ways stretching, condensing, speeding, warping, and looping linear
time and space, these perceptual abilities can easily seem to be talents
in accord with new realities instead of irrational delusions. In general,
the qualities praised fit perfectly with the kind of person frequently
described as highly desirable in corporate America: always adapting,
scanning the environment, continuously changing in creative and innovative ways.

If there is an increasing demand for restless change and continuous development of the person at all times, in many realms, then
manic-depression might readily come to be regarded as normal—even
ideal—for the human condition under these historically specific circumstances.

I do not mean to imply that appreciation for manic-depressives is
entirely new. In her book *The Psychiatric Persuasion,* Elizabeth Lunbeck has shown that in the early twentieth century psychiatrists generally reacted favorably to patients they diagnosed manic-depressive. As
psychiatrists of the time told it: "the story of the manic-depressive was
lively, often raucously so, and entertaining. Such individuals engaged
the world around them head-on, often wreaking havoc at home and
driving their relations mad . . . but they did so with a verve that drew

admiration. Their signal characteristics—loquacity, excitability, intense sociability, and mordant wit—differed from those of normal individuals only by virtue of their excess."[24] E. E. Southard, psychiatrist at the Boston Psychopathic Hospital, supposed that "every one of us is of manic-depressive stripe."[25] One difference I am seeing today is the extent to which the category of mania and of manic-depression has been widely taken up outside clinical settings, in many domains of popular culture. Another difference is the way manic-depression is associated with gender categories. In the early twentieth century, far more women were diagnosed with manic-depression than men, by some estimates twice as many. Lunbeck suggests that gender was encoded in the very category itself: "The most salient characteristics they saw in the manic patient were those associated in other contexts with an unbounded, out-of-control femininity that was at once frightening and alluring." Men diagnosed with manic-depression appeared "much like women" to their doctors, relatives, and friends: "excitable, distractable, and talkative, his conduct governed less by rational considerations than by plays of fancy."[26]

In contrast, dementia praecox (later schizophrenia) was coded "male." Its stolidity, stupidness, and catatonia "were merely the extreme, pathological manifestations of men's naturally more stable nature, just as the periodicity that characterized the manic mimicked in a more marked form the natural periodicity of women."[27] Today the gender differences for manic-depression have disappeared: "manic-depressive illness . . . is equally prevalent across gender," while major depression, with its immobility and numbness, is more common among women, not men.[28] It would be dangerous to make too much of such statistics, and at this point in my research, I am only speculating, but I wonder if we are watching today the inception of the male manic, seen as potent and effective despite his instability and irrationality. Is it possible that the "female" energy and exaggerated spirit Le Bon so despised in crowds now appears to us as a resource we need to tap for survival?[29]

One of the main features of mood disorders is that they are commonly medicated by drugs such as lithium or Depakote, often described as "managers" of manic-depression.[30] Interestingly, lithium is said to be resisted by patients more ferociously than any other drug psychiatrists prescribe. Widespread informal consensus labels lithium the drug that elicits by far more "failure to comply" than any other. As one patient told me, "I'd rather stand in front of a moving train than tell my psychiatrist I am manic, because I know she will make me take more lithium."[31] People who are not manic-depressive cannot understand this resistance to lithium, which promises you can "be normal." "But if you have had stars at your feet and the rings of planets through your hands, [and] are used to sleeping only four or five hours a night . . . it is a very real adjustment to blend into a three-piece-suit schedule, which, while comfortable to many, is new, restrictive, seemingly less productive, and maddeningly less intoxicating."[32] As Jamison puts it, manic-depression "destroys the basis of rational thought." If lithium restores it, then it is highly significant that many patients who have experienced being "irrational" refuse lithium as a kind of agent of the rational/modern, despite the agonies the alternative can produce.

Does the shift in the value attached to manic-depression actually amount to redefinition of the category, or is it just a swapping of the valence from negative to positive? If this shift continues, will it happen with a sense that meanings are constructed and so are changeable, or will it happen with a sense that a deeper, unchanging reality is being revealed? At this point in my research, these have to remain open questions.

MANIA IS GROUNDED

Outside these contexts in which mania is pushed off its pins, so to speak, there are other domains in which mania is gaining concreteness, timelessness, and facticity by being attached to particular physical causes that lie behind and explain its presence. It is well known that the effort to identify material causes for manic-depression is proceeding actively on many fronts, focusing mostly on genes and neurotransmitters. The mainstream press tells us scientific research will soon turn up "real physical correlates of eccentric behaviors."[33]

A number of scholars have shown how profoundly physical entities like genes and the DNA they are made of affect both scientific and cultural concepts of life, the body, and the person: summed up in what Dorothy Nelkin and Susan Lindee call "the power of the gene."[34] For example, an ongoing study at Johns Hopkins Medical School is designed to identify the genes responsible for manic-depression. Psychiatric histories and blood will be collected from volunteers and their families back two generations. When I talked to the person who screens prospective participants, she talked about the blood.

I ask: What's the blood for?
    In order to look at the genetic material, what we're trying to do is to compare within and across families because we are collecting several

families, and try to find out what individuals have in common who are affected versus those who are not affected and to try to find out what is making one individual have the disorder while the other doesn't.

And that would be something in the blood?

Yes, yes. We compare the genetic information.

Has this never been found before?

What we are looking for is the gene. We actually haven't found the gene; what we have found is an area of a certain size that we think the gene is located within so we are collecting additional families to sort of narrow down that spot on the chromosome that we think may contain one of the genes for the disorder.[35]

Kay Jamison makes clear the double-edged character of knowledge such as this study might produce. If manic-depression could have its genetic location specified, might there not be a rush to cleanse the species of these abnormal individuals with their defective genes? She argues that the sterilization or killing of the mentally ill during the Third Reich and the sterilization of the mentally ill during the eugenicist movement in the United States both provide chilling precedents. Studies in both these regimes considered whether to target manic-depressives particularly aggressively, and both recommended against it because manic-depressives were disproportionately found in the professional and higher occupational classes.[36] Tacitly accepting both the primacy of genetic causation and the applicability of natural selection to human populations, Jamison argues that the genes involved in manic-depression confer advantages on both individual and society, though like sickle-cell anemia, the cost to particular individuals may be great. Rather than challenging the assumptions involved in claims of genetic causation, she simply asserts that the genes responsible are important to the evolutionary strategies of our species.

A recent issue of *Newsweek* features a photo of Robin Williams on the cover alongside the headline "Are we all a little crazy?"[37] The issue stresses that the view of mental health in psychiatry is moving toward a continuum model in which there are many intermediate states surrounding any pathological condition. Neuroscientist Robert Sapolsky is quoted: "geneticists will come to the aid of psychiatrists in this debate. . . . The idea of a continuum represents a major cognitive breakthrough for genetics. It suggests that a middling genetic load [of mental illness genes] gives you a personality disorder, a lighter one gives you a personality quirk and a still lighter one gives you mainstream America." One of the researchers at Johns Hopkins, Francis J.

McMahon, weighs in on the new evidence that at least five genes are associated with manic-depression: "It may be that if you have only one gene, . . . you might be more susceptible to mood elevations that let you meet deadlines through a burst of activity or lead your business team across the finish line. The gene might be over-represented among artists and creative types." The journalist points out that "Robin Williams has described his deep depressions; any fan can see his mania, but someone with all 5 manic-depressive genes might be too buffeted by mood swings to function." The natural selection argument is reiterated in the context of asserting that many "disorders" serve useful functions for society. "Genes associated with mental illness might, in fact, keep society supplied with the personality types it needs." A neuroscientist explains that people with schizotypal personality disorder, for example, gravitate toward solitary occupations. "They are the lighthouse keepers and fire tower rangers." The article concludes, "when everyone is crazy, no one will be."[38]

The link between the array of genes and the array of psychic conditions happens by means of translated codes. In his recent trade book called *A Mood Apart,* physician Peter Whybrow explains: "My suspicion is that we will discover not one but a variety of genetic variations, each of which can predispose an individual to bipolar illness, including the milder forms such as seasonal affective disorder or the hyperthymia of great achievement. Probably each variant will be responsible for a slightly unusual genetic code in the library of instructions that builds the machinery of the neuron."[39] It is as if the text of the mind were simply and directly read off the genes.

Earlier, I said it was not clear whether the change of valence attached to the category manic-depression entailed the notion that its meanings slip with their context. But it is clear that the sliding scale of genes lying behind manic-depression does not include any such slipperiness. If every genetic variant, from only one gene present to many, is thought to cause specific kinds of behavior, then the category itself is only being divided into finer parts; it is not being redefined by its context.[40]

There is a tension in the contemporary meanings of mania: on the one hand, since its value is shifting, there is a sense in which its meaning seems indeterminate, constructed by the social context of our times. On the other hand, mania as a product of real physical causes seems increasingly a concrete and timeless fact. In my ongoing project, my purpose is not to dissipate this tension. Rather I want to shed light on why the manic has become such an active site of cultural attention in both these arenas.

THE CATEGORY "MANIA" IN ITS CONTEXT

What is there about mania that makes it a candidate for abnormality in the first place? The concept entails a degree of madness, loss of reason, or irrationality that is explicit in psychiatric definitions. In a manic state, the person does too much talking and assaulting, is overactive mentally and physically, has inadequate focus and garbled, delusional, or inexplicable speech. But mania is not just extreme emotions—enthusiasm, hilarity, anger, and so on. Psychiatric case studies and films of manic states (including actors simulating mania for the purpose of training medical students to identify it and actors like Robin Williams harnessing mania for comic effect) make it plain that in a manic state, it is as if the person has a crowd of people inside him or her all at the helm at once. But rather than the person being split between different selves who calmly take turns, the manic parts seem to jostle each other wildly for control. Aggressive, mocking, conciliatory, humorous, challenging, bragging, suspicious, hostile, expansive, open, belligerent, rejecting, fearful—every imaginable stance is enacted. Fear of the irrational set loose by a crowd is an old theme in Western history: at the turn of the century Le Bon laid it out: "Isolated, a person may be a cultivated individual; in a crowd, he is a barbarian— that is, a creature acting by instinct. He possesses the spontaneity, the violence, the ferocity, and also the enthusiasm and heroism of primitive beings." Of course, Le Bon participated in the Eurocentrism and racism of his day in likening crowd behavior to "primitive" behavior. In keeping with his time, he saw the irrationality apparently unleashed by crowds as characteristics of "women, savages, and children": "impulsiveness, irritability, incapacity to reason, the absence of judgment and of the critical spirit, the exaggeration of the sentiments . . . are almost always observed in beings belonging to inferior forms of evolution."[41]

There is certainly a high degree of suggestibility in all forms of mania, but perhaps, interpreted as exhibiting a *style* of speech behavior, a person with a crowd inside would seem less bizarre. In Robin Williams's stand-up comic routines, there is always some connection among the things said, but it is made through a kind of slipping: a word leads to an association, an object is put to every possible use, everything is pushed to yield its meaning. Robin Williams's comic style is part of a tradition that includes Jonathan Winters and Don Rickles; there is no doubt it is something worked for through discipline and training. Mania may not in any way be a result of training in this sense, but I would still suggest that we could say, after Jakobson, "Every mania is a style."

One way to understand the attention being paid to mania is to see it as a "colonizing" or "cultivating" of the irrational—the "culture" of mania as if the mind were a petri dish. If so the cultivating is probably related to certain ways commonsense conceptions of rationality are changing. The bounded, unitary, singular self seems to be giving way, allowing various kinds of separation within the self to be considered "normal" and sometimes desirable. These separations include dissociation of the self into parts, each of which has some autonomy from the others; experiencing the body as an object; experiencing the body as an object that shifts its dimensions in time and space.

Another possible reason for the colonization of the irrational in marginal people is that certain forms of irrationality that are in fact an intrinsic part of daily life in late capitalism are emerging into visibility. Perhaps the irrationality of the market and what you have to do and be to succeed in it are being more openly recognized. To be good at playing this crazy game, it helps to be a little crazy. "Rational choice" contains within it "irrational" impulses and desires. From this vantage point, we can look at the excess in Ted Turner as capitalist and as yacht captain (depicted and experienced as talent) as signs of apprehension: greater knowledge of what capitalism entails and greater fear of what it may require of people. A book that analyzes the workings of late capital is even called *One World Ready or Not: The Manic Logic of Global Capitalism.*[42] It is filled with references to "manic capital," oscillating with depression, and the calamitous consequences of both. The drop and rapid recovery of the stock exchange in October 1997 inspired a flurry of domain-crossing remarks in the business news, such as "If Wall Street were a person we'd think he was mentally ill."[43] Similarly, during the market swings of late summer 1998, a story from the *New York Times* business section was entitled "A Manic-Depressive Market Befuddles Even the Professionals": "Moodiness like this used to be an occasional thing in American stocks. Now, going from euphoria to depression is de rigueur. The trouble is, these market gyrations have consequences: they are highly distressing to investors' psyches and exceedingly costly to their portfolios."[44]

For mania to be colonized this way—for it to become an asset for Wall Street and its employees—it must seem stable and real. Abbott Labs would ask nothing less before investing in drugs to dampen mania in some people or heighten it in others. But simultaneously mania must shift its meaning to somehow traverse the huge divide between the irrational and rational, either making the irrational come to seem rational or forging a new category of sense somewhere in between.

Out of this dual movement arises the tension I have begun to lay out between *mania, the style* and *mania, the genetic condition.*

CLOSING REMARKS

The feminist paradigm has had a powerful effect on scholars in science studies far outside the traditional boundaries of "feminism." As one example, a conference in 1993 focused on "cyborg anthropology," a field that examines the argument that "human subjects and subjectivity are crucially as much a function of machines, their relations, and their information transfers as they are machine producers and operators."[45] Although the papers presented were on topics as diverse as PET scans, the human genome project, Prozac, the immune system, high-energy physics, and engineering education, all participants were basically in agreement that the field would weld together cultural anthropology, studies of science and technology, and feminism.[46] Feminism was central because it would be able to "stimulate widespread awareness of many different sources of stratification and hierarchy within science and technology, including race, class, and ethnicity. . . . feminist studies have detailed the ways in which science's claims to neutrality effectively generate hierarchies of power and control. [Feminism] . . . will insure that Cyborg Anthropology stays attuned to the diverse sources and forms of power constituted through science and technology and to alternative methodological strategies for providing analytical and critical understanding."[47]

Feminism has bequeathed to anthropological and cultural studies of science the insight that efforts to understand the production of knowledge need to be situated at junctures between scientific and technological research settings, on the one hand, and related cultural settings that reveal changing concepts of self and society, on the other. Extended to the social sciences, and to fields like the sciences of the mind that straddle the natural/social science divide, this approach can hope to illuminate the lines of power, stretching from one edge of a cultural field to the other, that are making something like mania seem at once more real, fixed, and known and more fictional, unstable, and unknowable.

NOTES

The research described here was made possible by a grant from the Spencer Foundation and an award from the NSF Cultural Anthropology Program. The data presented are solely the responsibility of the author.

1. Carol MacCormack and Marilyn Strathern, *Nature, Culture and Gender* (Cambridge: Cambridge University Press, 1980).

2. Toni Morrison, *The Nobel Lecture in Literature, 1993* (New York: Knopf, 1994), vi–viii.

3. Hugh Gusterson, *Nuclear Rites: A Weapons Laboratory at the End of the Cold War* (Berkeley and Los Angeles: University of California Press, 1996); Ashis Nandy, *The Intimate Enemy: Loss and Recovery of Self under Colonialism* (Delhi: Oxford University Press, 1983).

4. Avery F. Gordon, *Ghostly Matters: Haunting and the Sociological Imagination* (Minneapolis: University of Minnesota Press, 1997), 5–13.

5. "Given the prevailing binarized or dichotomized categories governing Western reason and the privilege accorded to one term over the other in binary pairs (mind over body, culture over nature, self over other, reason over passions, and so on), it is necessary to examine the subordinated, negative, or excluded term, *body* as the unacknowledged condition of the dominant term, *reason*. . . . The body has been and still is closely associated with women and the feminine, whereas the mind remains connected to men and the masculine" (Elizabeth Grosz, *Volatile Bodies: Toward a Corporeal Feminism* [Bloomington: Indiana University Press, 1994], 195).

6. Gordon, *Ghostly Matters,* 11, 8.

7. Ibid., 11.

8. For a recent representative collection of papers, see Floyd R. Davis and Joseph Dumit, eds., *Cyborg Babies: From Techno-Sex to Techno-Tots* (New York: Routledge, 1998).

9. For ethnographies of the production of knowledge in contemporary science in relation to cultural contexts, see Gusterson, *Nuclear Rites;* Sharon Traweek, *Beamtimes and Lifetimes: The World of High Energy Physics* (Cambridge, MA: Harvard University Press, 1988); Steven Epstein, *Impure Science: AIDS, Activism, and the Politics of Knowledge* (Berkeley and Los Angeles: University of California Press, 1996); and Paul Rabinow, *Making PCR: A Story of Biotechnology* (Chicago: University of Chicago Press, 1996).

10. See Emily Martin, *Flexible Bodies: Tracking Immunity in America from the Days of Polio to the Age of AIDS* (Boston: Beacon Press, 1994); Sherry Turkle, *Life on the Screen: Identity in the Age of the Internet* (New York: Simon and Schuster, 1995); and N. Katherine Hayles, *How We Became Posthuman: Virtual Bodies in Cybernetics, Literature, and Informatics* (Chicago: University of Chicago Press, 1999).

11. Peter Kramer, "Why Someone Would Risk It All," *U.S. News and World Report* 124, no. 5 (1998): 40; Peter Kramer, *Listening to Prozac* (New York: Viking, 1993).

12. Peter C. Whybrow, *A Mood Apart: Depression, Mania, and Other Afflictions of the Self* (New York: Basic Books, 1997), 109.

13. Ibid., xix.

14. Ibid., 49.

15. Roman Jakobson, *On Language* (Cambridge, MA: Harvard University Press, 1990), 130.

16. Abigail Padgett, *A Child of Silence* (New York: Mysterious Press, 1993); Tim Willocks, *Green River Rising* (London: J. Cape, 1994); Kaye Gibbons, *Sights Unseen* (New York: G. P. Putnam's Sons, 1995); Kay R. Jamison, *An Unquiet Mind* (New York: Knopf, 1995).

17. Jamison, *Unquiet Mind.* Most people I met in support groups during fieldwork felt they had little in common with the tormented, but creative, geniuses of Jamison's account. What is interesting to me about the reappearance of the old theme of the mad genius in this context is the implication that a certain amount of madness is good for, even essential for, everyone's survival in the contemporary world.

18. Ibid., 3.

19. Ibid., 105.
20. Ibid., 267–70.
21. Michael A. Bernstein, *A Bitter Carnival: Ressentiment and the Abject Hero* (Princeton, NJ: Princeton University Press, 1992), 7–8.
22. Jamison, *Unquiet Mind*, 80.
23. Frederick K. Goodwin and Kay R. Jamison, *Manic-Depressive Illness* (New York: Oxford University Press, 1990), 23.
24. Elizabeth Lunbeck, *The Psychiatric Persuasion: Knowledge, Gender, and Power in Modern America* (Princeton, NJ: Princeton University Press, 1994), 146.
25. Ibid.
26. Ibid., 149.
27. Ibid., 150.
28. Goodwin and Jamison, *Manic-Depressive Illness*, 168.
29. Gustave Le Bon, *The Crowd, a Study of the Popular Mind* (1895; reprint, Marietta, GA: Cherokee Publishing Co., 1982). For more on Le Bon's gendered imagery, see M. Norton Wise, "Time Gendered and Time Discovered in Victorian Science and Culture," in *From Energy to Information,* ed. Linda Henderson and Bruce Clark (Stanford, CA: Stanford University Press, forthcoming).
30. Depakote recently surpassed lithium to become the most prescribed drug to treat manic-depression. See G. Gunset, "Mania Drug Is 'Most Prescribed,'" *Chicago Tribune,* June 2, 1998, sec. 3, p. 1, col. 1.
31. Patient narrative, collected by author.
32. Jamison, *Unquiet Mind*, 91, 92.
33. Sharon Begley, "Is Everybody Crazy?" *Newsweek,* Jan. 26, 1998, 50–56. In a medical text on mania, Belmaker and van Pragg make the point that whatever role each factor is finally given, we are zeroing in on things that are physical causes. To describe both the certainty of what is known and the uncertainty that remains, they suggest an analogy: we know manic-depression is a fruit but "we don't know whether it's a citrus that will divide itself into separable sections or an apple that we must divide along arbitrary lines" (R. H. Belmaker and H. M. van Pragg, "Mania: Disease Entity or Symptom Cluster?" in *Mania: An Evolving Concept,* ed. R. H. Belmaker and H. M. van Pragg [Jamaica, NY: Spectrum Publications, 1980], 1–5).
34. Dorothy Nelkin and M. Susan Lindee, *The DNA Mystique: The Gene as a Cultural Icon* (New York: W. H. Freeman, 1995).
35. Interview conducted on Nov. 2, 1998, at Johns Hopkins Medical School.
36. Kay R. Jamison, *Touched with Fire: Manic-Depressive Illness and the Artistic Temperament* (New York: Free Press, 1993), 255.
37. *Newsweek,* Jan. 26, 1998. The story went by a slightly different headline: Sharon Begley, "Is Everybody Crazy?"
38. Ibid., 53, 55.
39. Whybrow, *A Mood Apart,* 162–63.
40. In Bahktin's terms, the language of genetics is referential: it is "direct," "unmediated," and "referentially oriented" in that it "recognizes only itself and its object, to which it strives to be maximally adequate. . . . Speakers of 'direct, unmediated discourse' . . . do not take into account the already-spoken-about quality of the object or, at least, not in a way that implicitly challenges the authority of their own speech. . . . they simply name their referent" (Gary S. Morson and Caryl Emerson, *Mikhail Bakhtin: Creation of a Prosaics* [Stanford, CA: Stanford University Press, 1990], 148).
41. Le Bon, *The Crowd,* 12, 16.
42. William Greider, *One World, Ready or Not: The Manic Logic of Global Capitalism* (New York: Simon and Schuster, 1997).
43. Louis Uchitelle, "Confusion as an Economic Indicator," *New York Times,* Nov. 2, 1997, sec. 4, p. 1, col. 1.

44. Gretchen Morgenson, "A Manic-Depressive Market Befuddles Even the Professionals," *New York Times,* Sept. 11, 1998, sec. 2, p. 1, col. 2.

45. Gary Lee Downey, Joseph Dumit, and Sharon Traweek, unpublished conference materials, precirculated to participants prior to "Cyborg Anthropology," held at the School of American Research, Santa Fe, NM, Oct. 1993.

46. The conference papers were subsequently published in Gary Lee Downey and Joseph Dumit, eds., *Cyborgs and Citadels: Anthropological Interventions in Emerging Sciences and Technologies* (Santa Fe, NM: School of American Research Press, 1997).

47. Downey, Dumit, and Traweek, "Cyborg Anthropology" precirculated conference materials, 4–5.

THIRTEEN

Gendering the Epidemic: Feminism and the Epidemic of HIV/AIDS in the United States, 1981-1999

EVELYNN M. HAMMONDS

In 1981, physicians across the United States reported that an unusual number of healthy young men, the majority of whom were homosexual, were falling ill from a then-mysterious disease. Many of them suffered from Kaposi's sarcoma, a rare cancer that was usually found only in elderly Italian and Jewish men and some Africans. Some were also suffering from *Pneumocystis carinii* pneumonia (PCP), an infectious lung disease that was also rare. By the end of the year the Centers for Disease Control (CDC) had identified these diseases as part of a new disease entity, which was called gay related immune disease (GRID). By the end of 1981 over two hundred cases of the new disease had been reported and the CDC officially named the disease the acquired immune deficiency syndrome (AIDS). In that same year, AIDS was also declared to be epidemic in the United States. The disease was linked almost immediately to gay men, though buried in the first CDC report was also the first identified case of a woman infected with the disease. By the end of the year five cases in women were reported. In 1983, the virus associated with AIDS, a human retrovirus (HIV), was isolated. In that same year it was clear that HIV/AIDS was now a pandemic, spreading rapidly around the world.

Over the next decade the numbers of women infected and dying from AIDS slowly increased along with increasing numbers of infants infected in utero or shortly after birth. In 1981 women accounted for 3 percent of AIDS cases; that number doubled to 6 percent in 1984 and more than tripled to 19 percent in 1995, "the highest proportion yet reported among women."[1] Women of color, specifically African American and Latina women, accounted for more than three-fourths of all AIDS cases in women: by 1995, 55.8 percent were African Amer-

ican and 20.7 percent were Latina, although African American women
constitute only 15 percent of the U.S. population and Latinas only 7
percent. HIV/AIDS is now the leading cause of death among African
American women aged twenty-five to forty-four, and the death rate
from the disease in this group is nine times higher than that of white
women in the same age group.[2] At the end of 1995, a total of 71,818
HIV-positive women had been reported to the CDC.[3]

The AIDS epidemic has been profoundly gendered. The figure of
the homosexual male has been inextricably linked to the disease in the
popular imagination and in the medical and public health literature.
The fate of women in the epidemic, however, has been difficult to characterize from the beginning. Feminists flush with the success and hard-
won expertise gained in the women's health movement, arguably one
of the most successful aspects of second-wave feminism in the United
States, were in turn baffled and angry at the lack of medical, public
health and popular press response to the plight of women infected with
AIDS. In this chapter, I want to analyze the strategies employed by
feminists in dealing with the gender issues at the center of the AIDS
epidemic. These strategies, I will argue, bring to light a central problematic in contemporary feminism: how to deal with the fragmentation
of the category "woman" in theory and in practice. Or to state it differently, how can mainstream feminism address, speak, and act in the
name of women who are socially marked and marginalized in different
ways by race, class, sexuality, *and* disease? I begin with the many
"women" in the AIDS epidemic followed by a discussion of feminist
strategies: from identifying the "women" and making "women" visible to moving "women" from the margin to the center of discourse,
policies, and practices in the epidemic.

Before I begin let me say a word about the methodological approach
I take with respect to the AIDS epidemic in this chapter. Historians
of medicine have rightfully argued that the epidemic of HIV/AIDS is a
biological and a cultural phenomenon.[4] Disease is socially constructed,
created by people making meaning out of biological phenomena. Biological mechanisms, the historian of medicine Charles Rosenberg has
argued, define and constrain social response. AIDS is a profoundly
socially mediated epidemic. Our response to the virus has been shaped
by the tools of modern virology and immunology that made it visible;
the epidemiological categories used to define its routes of transmission
and its victims; the language and representations used to describe the
experience of the sick and the dying; and the deeply felt social attitudes
about sex, drug abuse, and disease HIV/AIDS exposes in our society.
I take it in this essay that all these factors simultaneously serve to create

what AIDS *is*. What follows then is an overview of some of the questions that women have faced in the AIDS epidemic.

THE "WOMEN" IN THE EPIDEMIC

In the first decade of the AIDS epidemic, women could be placed in two categories: the *infected* and the *affected*.

The Infected Women

Within months of the identification of the disease, the infected women represented an anomaly. They were defined by what they were not: they were not male and not homosexual. Later, as the disease spread to other groups, the CDC instituted its policy of defining the groups at risk for AIDS, the so-called 4-H club: homosexuals, Haitians, heroin (and other injection drug) users, and hemophiliacs. This list firmly established the view that the disease was associated with immorality, drug abuse, and deviant sexual behaviors. The public representation of these risk groups split the population into the "Others," those deviants who contracted AIDS through immoral behavior, and the "innocent" victims, who contracted the disease through blood transfusions. The stigmatization that the risk group designations implied was not lost on those characterized this way. Haitian activists immediately protested the stereotyping of their peoples. The removal of Haitians from the CDC risk groups only served to harden the view that homosexuals as a group were really the only people at risk for AIDS. All research and public health policies were directed at this population at this time. Of course, this made sense because homosexual men did make up the largest number of AIDS cases and deaths during this decade. Yet the number of cases and deaths among women was rising, and none of the research on the disease was oriented toward understanding women's risk or the natural history of the disease in women.

The little that was known about these women was largely gleaned from reports from women's prisons, drug rehabilitation centers, and the emergency wards of public hospitals. The women physicians treating these women were initially surprised that women were dying of AIDS. Others in the field of public health were worried, recognizing that because AIDS could be transmitted sexually, there could be little doubt that women would be infected. After all, there was no known sexually transmitted disease that affected only one sex. In the medical and public health literature, however, no full-scale analysis of women and AIDS appeared. In CDC reports at this time, the category "female" did not yet appear in epidemiological accounts. In early 1987, evidence of the virus was found among "prostitutes." Yet this category was never defined explicitly in terms of behaviors that put such women at risk. Instead, researchers drew upon outmoded and ill-informed stereotypes about what such women did. Most significantly, they were defined by the risk they posed to men rather than the risk the virus posed for them.[5]

Although little was known about the early cases among women, the media and policymakers suggested that women who acquired the "gay" plague were themselves sexually deviant. Some argued that little attention was paid to AIDS in women because the number of cases in women was considered too small for quantitative analysis. Race was a factor as well. The largest percentage of the cases in women involved women of color, both African American and Latina. Before 1987, virtually no attention was paid to the fact that AIDS had spread into communities of color.

If one tried to write a history of women in the AIDS epidemic using health department reports, medical and scientific journal articles, or newspaper reports up through 1987, it would be easy to document that infected women were invisible in the epidemic. The historian might also be tempted to characterize AIDS in women as completely separate and distinct from AIDS in men, not only in the biology— the differing natural course of HIV in female bodies and the different opportunistic infections that appeared in women—but also in representations of AIDS in popular culture and public policy. Following this track could lead to viewing the epidemic in a linear fashion, slowly and inexorably traveling from one group, gay men, to women, to communities of color; from the Northeast to the South; and reaching peaks at different moments in time. Yet, I would suggest that this approach would be limited because it would fail to seriously consider the consequences of human actions in the spread of the virus. It would fail to represent how the connections between women and men variously located within many communities would remain factors in our understanding of this epidemic.

By the late 1980s, the women in the AIDS epidemic were categorized by the route through which they were exposed: as sexual partners of infected men, as sharers of needles with infected men, as prostitutes, or as bad mothers for infecting their children. In this period when stigma against people with AIDS was high, these women were perceived as the "Others."

For women infected with AIDS, blame and stigma were a constant part of their lives. The question, How did you get it? was repeatedly asked. One woman noted, "If the answer was drugs or needle sharing, other women listening for the answer would sigh with relief . . . 'it couldn't be me.' " For infected women, revealing their HIV status had many consequences, including rejection by family and friends and possible loss of employment. For women of color, discrimination was nothing new. AIDS just highlighted the long-standing racism they had experienced with the health care system and American society more generally. As one of the first staff members in the New York City Department of Health AIDS unit put it, "The minority community knew about these problems, but the gay community wasn't bothered by them before. Now, when they are trying to access the system, they are raising hell, and the minorities are saying: 'Welcome to the deal.'"[6]

Despite their experience with discrimination, communities of color faced serious challenges in confronting AIDS in 1988. The church, community service organizations, and other traditional sites for addressing community issues could not deal with AIDS. The pervasive association among AIDS, gay men, and drug abuse evoked widespread and deeply felt religious strictures against homosexuality and drug abuse and thus hindered action on the epidemic by these institutions. Although organizations such as New York City's Gay Men's Health Crisis (GMHC) and Boston's AIDS Action Committee (AAC) formed because as one gay activist noted, "we couldn't wait for somebody else to help people with AIDS," they had little impact in communities of color. As a result, in 1985, activists of color in Boston formed the Multicultural Concerns Committee (MCC) within the AAC. African American attorney Paula Johnson, who served as cochairperson of the organization, noted that the AAC was perceived as a white gay organization and many people of color refused to use its services. As a result the MCC moved out into communities of color seeking an audience with almost anyone who would listen. They encountered barriers at every turn. The silence around AIDS in these communities was virtually impenetrable.[7] This silence contributed to making it almost impossible for women of color to know that they were at risk for AIDS. At the end of the first decade of the epidemic, within and outside their communities infected women confronted stigmatization, discrimination, abandonment, fear of death, and loss of family members. No national organization or campaign existed to make visible their plight.

Women Affected by HIV/AIDS

As more men became infected, the category of women "affected" by the epidemic grew as well. The focus on the so-called deviant infected women also contributed to overlooking the many other ways that women were being affected by the AIDS epidemic: as wives, mothers, siblings, doctors, nurses, and social workers. These women experienced the epidemic in complicated ways because of their connections to infected men and women and to the medical, public health, and social welfare establishment. An analysis of oral histories and testimonies from women at this time indicates that women were affected from the very beginning of the epidemic. They faced, along with gay men, stigmatization, shame, and loss. Family members of persons with AIDS suffered greatly. One woman described her feelings as her husband died of AIDS as follows: "This is my disease, I think. . . . I don't have the virus but I have the disease. It's a disease of exposure. It's a disease that tells people about Tom's past and therefore mine. I feel tainted and ashamed."[8] A mother spoke of the response of her family to her son's death of AIDS in 1984: "My relatives can't deal with homosexuality . . . when I told my mother that Michael was sick, she just drove me crazy: 'How did you let him get sick?' It was as if it was my fault."[9] In their traditional roles as caretakers of sick family members women inadvertently found themselves on the front lines of the epidemic.

As physicians, nurses, and social workers, women also confronted the ravages of AIDS. A young physician working in a Bay area hospital wrote about the enormity of the disease: "Many of them were my age, and many were dying. Several of them had arrived at the county health care system through tragic personal circumstances, their AIDS diagnosis had cost them their jobs and sometimes their health insurance. I was overwhelmed by their illness, their very complicated medical problems, their awesome psychological and emotional needs, and their dying. I was frightened by the desperation of many who wanted to be made well again or to survive that which could not be survived."[10] Social workers were burning out under the stress of trying to save people with AIDS, especially the infected women. "We ask ourselves if we can do preventive work around AIDS with women who are still shooting. We see them in hospitals, temporary shelters, homes where they live with male partners or other mothers. We try to put ourselves in their places, but we cannot. . . . On my good days, I view it as a challenge. On my bad days, I wonder what will become of us. All of us. Women included."[11] By the late 1980s, it was clear to many women physicians, activists, social workers, and health care providers that women were increasingly at risk for AIDS.

Another group of women affected by the epidemic were those working in AIDS organizations. The male-dominated structure and focus of those organizations forced many of them to establish informal networks with other women in order to gain recognition of what was happening to women. Yet these women professionals working in AIDS organizations found it almost impossible to put women's issues on an agenda that was framed by and for gay men. Many of them had been active in the women's health movement and were used to confronting the medical establishment, yet they had little experience working with gay men. Meanwhile mainstream feminist organizations were suspicious of making coalitions with gay men, who they felt did not share their feminist politics. Ironically, AIDS organizing owed a great debt to the feminist movement of the 1970s. It picked up and carried forward from that movement a commitment to personal politics. Gay male activists adopted the way in which women challenged the medical system by questioning every aspect of medical and public health response to the epidemic. Yet despite this shared approach, gay men as a group had several advantages over women. "Gay men living with HIV, to their credit, leveraged their skills and their tenuous legitimacy, along with a very powerful claim to the experience of oppression at the hands of the medical system, to exercise some control over how they were collectively represented in relation to HIV."[12] As AIDS activist Cindy Patton has noted further, gay men had privilege, education, and greater access to doctors, analysts, media outlets, and researchers because they were men. More important, gay men as the *infected* and as the *affected* were able to represent themselves in the epidemic. For women the situation was quite different.

The women *infected* were largely from a different class, race, and educational background than those *affected* were. The women physicians, epidemiologists, social workers, and other professionals and activists among the affected women found it difficult to represent the needs of the poor, the drug abusers, and the women of color who made up the bulk of the women with AIDS. When those women tried to represent themselves, they faced almost insurmountable obstacles of class and race. Few of the women living with HIV had the cultural capital necessary to challenge the research and policy establishment on their own behalf. By the late 1980s, there was a very serious gap between infected women's needs and the activities of gay community–based AIDS organizations. The women caught in the middle were the feminist women professionals who wanted to help infected women.

Meanwhile, media reports continued to fragment the women in the epidemic by highlighting the risk of these women to others. Negative characterizations of bad mothers or bad wives dominated accounts of women with AIDS. In the minds of many, women were the vectors by which HIV was transmitted to men or to children. These bad women were overwhelmingly women of color who were depicted as responsible for their HIV infection. Conversely, heterosexual, white, middle-class women were more and more frequently viewed as the "innocent victims" of secretly bisexual or drug-using men in their lives. Mainstream discussions of AIDS and women focused on fragmenting women into categories of innocent and guilty; of women who counted as women, worthy of protection and support, and women who did not count as women, condemned to be policed and controlled by the state.[13]

MAKING WOMEN VISIBLE

If it is possible to characterize the diverse efforts of feminists on behalf of HIV-infected women in this period, one might say they employed a strategy of trying to make infected women visible. In order to make visible the plight of infected women they lobbied for very concrete things. They fought for funds to conduct clinical studies to demonstrate the effects of HIV in women. They lobbied to have the regulations governing clinical trials of new drugs to fight AIDS changed to include women. At international conferences they challenged the male leaders of the international AIDS community to conduct research on women, and most important, they clamored for a change in the CDC surveillance definition[14] of AIDS to include the gynecological infections that were uniquely manifest in women. They also formed their own organizations. The first was the Association of Women's AIDS Research and Education (AWARE), followed in 1986 by the Women and AIDS project and, one year later, the Women and AIDS Resource Network (WARN). These organizations were started with shoestring budgets to coordinate fund-raising and to collect and disseminate information about women and AIDS. Projects such as these represented the first organized efforts to concentrate on women infected with AIDS. They produced much needed educational materials, supported clinical and epidemiological studies on women, and offered the first support services for infected women.[15]

To make the infected women visible, these organizations attempted to demolish stereotypes about the women who were infected. They published studies showing that the women who used drugs were not the social deviants portrayed in the public health literature and the popular press. Indeed, many of these women had been victims of sexual abuse, depression, and crippling poverty. They were often pushed into prostitution by male drug dealers, becoming "virtual slaves to both drugs and the dealers."[16] Tackling head on the view that women were the "invisible pass through" to men, these feminists agitated for the consideration of women as whole human beings. They emphasized that it was a dangerous myth that only flagrantly "promiscuous" women were at risk for AIDS. This myth led many women who were not drug abusers or prostitutes to believe that as long as they were not using drugs or having intercourse with numerous men every day, they were safe. Writer Gena Corea, in her history of women and AIDS, *The Invisible Epidemic,* noted the fallacy embedded in the myth that "faithful" women were safe. "Many of the infected women were homemakers; the working poor; women whose partners' drug injection or sex practices—not their own—had been the source of HIV transmission."[17]

Yet as the women's AIDS organizations continued to hold public hearings before state legislatures and congressional committees and to lobby foundations for money for research on women with AIDS, they were only marginally successful at transforming the image of infected women in the epidemic.

MOVING WOMEN FROM MARGIN TO CENTER

In 1993, the CDC issued a revised surveillance definition of AIDS, recognizing for the first time clinical manifestations of HIV disease unique to women.[18] This change in the surveillance definition resulted in an immediate increase in the number of recognized cases of HIV in women. In the same year, the Food and Drug Administration (FDA) officially lifted its ban on women's participation in early experimental drug trials. As cases of AIDS in men declined, whether because of successful public education efforts or changes in behaviors, the cases in women continued to rise. In 1994, more than 14,000 new AIDS cases were diagnosed in women, representing 18 percent of all new cases and a dramatic increase in incidence since 1986. By 1997, women represented the fastest-growing "risk group" for acquiring HIV infection.[19]

Behind the change in the surveillance definition were the increasingly strident voices of women demanding recognition. In the mid-1990s, many of the women working on AIDS had begun to recognize that much more needed to be done for infected women. Gena Corea described the pent-up frustration, anger, and dismay expressed by women activists, health professionals, and infected women at the 1990 National Conference on Women and HIV Infection.[20] Though health professionals demanded more funding for clinical and epidemiological studies on women, this meeting made it abundantly clear that women health professionals were fed up with the barriers imposed by the CDC and other federal agencies in funding such studies. Health education workers demanded more research on spermicides and other drugs that could help women protect themselves against infection. The testimonials from infected women poignantly documented the pervasive lack of services for infected women and their children and families. Physicians noted that, a decade into the epidemic, obstetricians and gynecologists who provided most of the medical care for women still received little training on how to recognize HIV infection in women. And as a result, the majority of women presenting to these physicians had no idea that they were infected with HIV.

Corea ends her narrative on a despairing note with the description of this meeting. The challenges that this epidemic posed for all the women, infected and affected, seemed to require a renewed focus on activism for women, although Corea herself has little to say about how such activism would begin. By the mid-1990s, what is most evident in the various literatures about AIDS is the increasing fragmentation of approaches to addressing the problems of women with AIDS. The various actors in the epidemic were becoming isolated from one another. "Listening to the different voices and language of clinicians, MD's epidemiologists, psychologists, community-based organizers, AIDS activists, and people infected or affected by HIV/AIDS, it sometimes seems as though they were sequestered on separate planets and spoke entirely different languages," wrote the editors of a recent compilation of articles on the gender politics of HIV/AIDS in women.[21] These editors acknowledge the failure to find an integrated approach for women in the epidemic: "With their separate conferences, journals, and foci, members of simultaneously disparate and overlapping fields have faced very real impediments to staying current with one another's findings and developments. But the cost of this kind of ideological, professional, and institutional segregation is very high: ultimately it makes a multifaceted and fully informed response to the pandemic virtually impossible."[22]

The efforts to make visible the plight of women in the AIDS epidemic by these diverse groups of professionals and activists were uneven. Some progress was made on some fronts; funding for research and services had indeed improved in the 1990s. Yet feminist scholars increasingly grappled with the problem of women and AIDS during this period. Cindy Patton, an activist and scholar who has written some of the earliest and most cogent analyses of women and AIDS since the 1980s, defined the problem as one of "speaking for" rather than "speaking as."

To the extent that our sex is accepted (among ourselves and by sources of power) as a real line of affinity, we can "speak together" as different kinds of women, even if some of our voices are allowed to be more articulate or credible. The early stereotyping of HIV as a disease of immorality and deviance challenged women's assertion of ourselves as a sex-class.[23]

More important, she argued:

"Woman" is too many things in the epidemic: epidemiology's "partners of," communities' "other half," the "general public's" most vulnerable part, the dangerous sexual outlaws that prey on wayward men. They are either a demographic exception (not gay men) or the idealized case where we can see the unbiased (by sexuality) "truth" of the pathos of the epidemic. Simultaneously passive, innocent victim and monstrous, infectious sex organ, women do not yet have voices that give them purchase on the representational and medical systems that engulf them.[24]

Feminists like Patton joined with women health professionals and activists in becoming more adept at describing the problem of women in the AIDS epidemic. Such analyses have also been accompanied by certain silences. Cultural critic Paula Treichler noted the silence of the explicitly feminist media on the epidemic. In particular she noted that virtually no attention was paid to the AIDS epidemic in the women's studies literature between 1982 and 1986 and that fewer than ten articles were published by 1988. Only sporadic attention was paid in fields like cultural studies, art history and criticism, media studies, sociology, history of science, and cultural anthropology.[25] For others, AIDS had become an occasion for analyzing the dominant feminism of the 1970s and 1980s. In most cases AIDS in fact intensified questions many feminists had been asking since the late 1970s: When we say we are feminists, who are the "we" and at what price?[26]

CONCLUSION

What is surprising is not that AIDS became an occasion for asking such questions. Rather, it is the paucity of answers that is surprising and disturbing. As Treichler notes, feminist interventions into medical discourse had "altered medicine's linguistic and cultural constructions of women and reshaped medical practice in a number of ways—epistemological changes brought by a social movement."[27] Indeed, feminist analyses had moved from stories of the "great women doctors"; to highly polemical analyses of women's victimization at the hands of a male-dominated medical profession; to studies of how medicine defined and attempted to control women's lives; to more nuanced discussions of the complex and dynamic relationships between women (as patients and practitioners) and medicine (both practice and theory). These latter studies increasingly pointed to the need to understand the role of women's agency. Women had never been simply the passive recipients of medical advice and therapeutics. Nor had they been defined simply by male-constructed medical theories. As historian of medicine Rima Apple noted, women "are active participants who at times resist, at times embrace, and at times create the conditions in which we find ourselves."[28] Indeed, the current sophistication in feminist studies of medicine and science is in part due to the ways in which feminist theory and activism intersected. Yet, while feminist scholarship has made great strides in writing about some women, it has told us little about the lives of many women, including those who are illiterate, working class, immigrant, and women of color.

Treichler's argument that feminist omissions and silences surrounding AIDS were produced out of a complex nexus of tangled loyalties and stereotypes involving "sexuality, sexual orientation, race and class issues" is undoubtedly correct as far as it goes.[29] Yet her analysis and those of Patton and other feminists who have long studied the epidemic have never fully grappled with the problem of race and class in the epidemic. Simply identifying these as problematic aspects of the category "women" accomplishes little. Nor does it help to frame analyses of the epidemic if these two issues are not at the center of feminist analyses.

Gendering the AIDS epidemic requires more than moving women from the margins to the center of attention. Feminists need to take up a more complicated notion of gender. Gender theorists like Joan Wallach Scott have emphasized that we must examine the ways in which gendered identities are substantively constructed and relate our findings to a range of activities, social organizations, and historically specific cultural representations.[30] Additionally, in societies such as ours that are structured around racial and class divisions, these concepts must be considered alongside gender as key markers of power relations. Investigating how medicine and public health structure power relations that construct gendered, "raced," and "classed" identities must be at the center of our work.

Ultimately, gendering the AIDS epidemic should make visible not just problematic categorizations of women but also other missing elements that these categorizations have made invisible. The behaviors of heterosexual and bisexual men have been largely ignored in feminist analyses of the epidemic, yet it is the behaviors of these men that have contributed to the growing risk of infection in women. The AIDS epi-

demic has revealed that heterosexual relationships must be reconstructed in the age of AIDS in order for women to reduce their risk.
Women of color as figured in AIDS narratives have consistently been
the site where whites' fears of disease are banished. The tragedy is that
what is obscured by accepting the categorization of women of color
as "Other" is the fact that the AIDS epidemic is becoming the great
leveler of women. "Fear of becoming infected and an inability to ensure that they remain uninfected are common to virtually all heterosex-
ually active women, irrespective of their race, their education, their
social class, their lifestyles, their marital status, their legal rights, or
any other socioeconomic variable." [31]

Gendering the AIDS epidemic involves the production of analyses
that resist and reframe the fragmentation of the category of woman
along class and racial lines. Simply put, "fragmentation" itself must
be theorized. In my own work on African American women and AIDS
I have suggested that feminist analyses of women and AIDS need to
express how gender and race are related in order to disrupt the stereotypes of both, if we are to produce effective and meaningful responses
for all women. [32] A first step would be to acknowledge how our understanding of race in the United States is profoundly dependent upon
the visual.

One of the most provocative examples of a representation of
women and AIDS that embodies the complex history of representations of African American women and also speaks to all women is
provided by the artist Lorna Simpson. Simpson's photograph for the
Art against AIDS Project depicts the torso of a dark-skinned woman
dressed in a white sheath. Her arms are crossed just below barely dis-
cernible breasts. In the way she holds herself the figure seems to be
vulnerable and yet also determined. There is no face. The body speaks
the text written upon it:

> a lie is not a shelter
> discrimination
> is not protection
> isolation is not a remedy
> a promise is not a prophylactic

Simpson's photograph is a critique of the prevailing representations
of African American women's bodies, and of women and AIDS. The
photograph is both specific and universal: the body of a dark-skinned
woman—she could be African American or Latina—is dressed in a
sheath that is similar to the gown any woman would wear in a medical

setting. It is the text that disrupts dominant representations of women's risk for AIDS. Rather than speaking *for* women of color with AIDS, the photograph speaks *to* the questions AIDS raises for any woman: the potential infidelity of partners; the discrimination that women encounter in the health care system; the silence that accompanies the fear of loss of community and family; and the too-often-futile hope that "love" can protect women from AIDS. And finally the figure's embrace of herself can be read as a call for self-empowerment.

Simpson's photograph evokes race and disrupts racial stereotyping at the same time. It speaks to women of color, not for them. It speaks to all women from the perspective of a woman of color and thus highlights both the subjective and the universal. Simpson successfully and powerfully disrupts the negative representations of women of color by her skilled insistence that representations must be challenged by articulation. She encourages us to recognize that what happens on the level of representation mirrors what occurs at the level of AIDS prevention. Ultimately it is *articulation*, not *representation*, that feminists have to offer the women in the AIDS epidemic. Gendering the AIDS epidemic and empowering women is then finally a challenge to feminist ethics. In the end, resisting fragmentation and producing effective and empowering identities that all women can claim in the midst of this epidemic can save women's lives.

NOTES

1. Nancy Goldstein and Jennifer L. Manlowe, eds., *The Gender Politics of HIV/AIDS in Women: Perspectives on the Pandemic in the United States* (New York: New York University Press, 1997), 9.
2. Ibid.
3. Ibid. The picture of HIV/AIDS among women is startling, as Goldstein and Manlowe note: the median age was thirty-six years, and 90 percent of cases were reported in women twenty to fifty years of age. Twenty-five percent of women were diagnosed with AIDS between the ages of twenty and thirty years, indicating that substantial transmission of HIV occurred during adolescence and the early twenties. Women accounted for the majority of cases infected through heterosexual contact or injection drug use.
4. For a discussion of the social construction of disease, see Elizabeth Fee and Daniel Fox, eds., *AIDS: The Burdens of History* (Berkeley and Los Angeles: University of California Press, 1988), especially Charles E. Rosenberg's chapter, "Disease and Social Order in America: Perceptions and Expectations," 12–32; and Elizabeth Fee and Daniel Fox, eds., *AIDS: The Making of a Chronic Disease* (Berkeley and Los Angeles: University of California Press, 1992).
5. Indeed, subsequent research would show that heterosexual transmission was five to six times more efficient from a man to a woman. Women were much more likely to be infected by men than vice versa. Gena Corea, *The Invisible*

Epidemic: The Story of Women and AIDS (New York: Harper Collins, 1992), 222.

6. Sunny Rumsey, "Communities under Siege," in *AIDS: The Women,* ed. Ines Rieder and Patricia Ruppelt (San Francisco: Cleis Press, 1988), 190.

7. Paula C. Johnson, "Silence Equals Death: The Response to AIDS within Communities of Color," *University of Illinois Law Review* 1992 (1992): 1075–83.

8. Rieder and Ruppelt, *AIDS,* 20.

9. Ibid., 32–33.

10. Ibid., 101.

11. Ibid., 103.

12. Cindy Patton, introduction to Nancy L. Roth and Katie Hogan, eds., *Gendered Epidemic: Representations of Women in the Age of AIDS* (New York: Routledge, 1998), x.

13. In this period when methods to control the spread of AIDS were discussed, repressive methods considered included quarantine, automatic testing of women in prenatal clinics, mandatory testing of prostitutes, installation of electronic devices for surveillance of persons with AIDS. Such efforts succeeded on those occasions when women and sexist stereotypes of women were used to facilitate and justify them. Ibid.; and Corea, *Invisible Epidemic,* 34.

14. A "surveillance definition" is the definition used by the CDC to monitor incidence and mortality rates of specific diseases.

15. Corea, *Invisible Epidemic,* 58–60.

16. Ibid., 66–67.

17. Ibid., 86.

18. Centers for Disease Control, "1993 Revised Classification System for HIV Infection and Expanded Surveillance Case Definition for AIDS among Adolescents and Adults," *Morbidity and Mortality Weekly Report* 41, RR17 (1992): 1–19.

19. Bill Rodriguez, "Biomedical Models of HIV in Women," in Goldstein and Manlowe, *Gender Politics,* 25.

20. Corea, *Invisible Epidemic,* 281.

21. Goldstein and Manlowe, introduction to *Gender Politics,* 10–11.

22. Ibid., 11.

23. Roth and Hogan, *Gendered Epidemic,* xiii.

24. Ibid.

25. Paula Treichler and Catherine Warren, "Maybe Next Year: Feminist Silence and the AIDS Epidemic," in Roth and Hogan, *Gendered Epidemic,* 134.

26. Ibid.

27. Ibid., 139.

28. Rima D. Apple, introduction to *Women, Health, and Medicine in America: A Historical Handbook,* ed. Rima D. Apple (New York: Garland Press, 1990), xiii.

29. Treichler and Warren, "Maybe Next Year," 141.

30. Joan Wallach Scott, "Gender: A Useful Category for Historical Analysis," in *Feminism and History,* ed. Joan Wallach Scott (Oxford: Oxford University Press, 1996), 169.

31. Elizabeth Reid, "Population and Development Issues: The Linkages to HIV and Women," in *Defining the Women and AIDS Agenda* (Arlington, VA: AIDSCAP Women's Initiative, 1995), 27.

32. See a fuller discussion of this point and the Simpson work in my "Seeing AIDS: Race, Gender, and Representation," in Goldstein and Manlowe, *Gender Politics,* 113–26.

CONTRIBUTORS

RUTH SCHWARTZ COWAN is professor of history and chair of the Honors College at the State University of New York in Stony Brook. Her current research focuses on the history of diagnosis, treatment, and prevention of genetic diseases. In 1997 she published *A Social History of American Technology* (Oxford University Press).

ANGELA N. H. CREAGER is associate professor in the Department of History and the Program in History of Science at Princeton University, offering courses on the history of biology and on gender and science. She publishes on the history of biomedical research, most recently in *The Life of a Virus: Tobacco Mosaic Virus as an Experimental Model, 1930–1965* (University of Chicago Press, 2002).

LINDA MARIE FEDIGAN is professor of anthropology at the University of Alberta in Canada. She has conducted fieldwork in St. Kitts, Guatemala, Costa Rica, and Japan. In addition to publishing on gender and science, she has written several books and many articles on primate life histories and relations between the sexes. Her most recent book, coedited with Shirley Strum, is entitled *Primate Encounters: Models of Science, Gender, and Society* (University of Chicago Press, 2000).

SCOTT F. GILBERT is professor of biology at Swarthmore College. He received his Ph.D. in biology and his M.A. in the history of biology from Johns Hopkins University. He teaches classes in embryology, developmental genetics, and the history and critiques of biology. He is presently chair of the Division of Developmental and Cell Biology of the Society for Integrative and Comparative Biology. He was recently awarded a John Simon Guggenheim Memorial Fellowship. He is the author of *Developmental Biology* (Sinauer, 2000).

EVELYNN M. HAMMONDS is associate professor of the history of science in the Program in Science, Technology, and Society at Massachusetts Institute of Technology. Author of *Childhood's Deadly Scourge: The Campaign to Control Diphtheria in New York City, 1880–1930* (Johns Hopkins University Press, 1999), she is currently writing a new book, *The Logic of Difference: A History of Race in Science and Medicine in the United States* (University of North Carolina Press, forthcoming).

EVELYN FOX KELLER received her Ph.D. in theoretical physics at Harvard University, worked for a number of years at the interface of physics and biology, and is now professor of history and philosophy of science in the Program in Science, Technology, and Society at Massachusetts Institute of Technology. She is the author of *A Feeling for the Organism: The Life and Work of Barbara McClintock* (W. H. Freeman, 1983); *Reflections on Gender and Science* (Yale University Press, 1985); *Secrets of Life, Secrets of Death: Essays on Language, Gender, and Science* (Routledge, 1992); *Refiguring Life: Metaphors of Twentieth Century Biology* (Columbia University Press, 1995); and *The Century of the Gene* (Harvard University Press, 2000). Her next book, *Making Sense of Life: Models and Explanation in Developmental Biology*, is expected to appear by the end of 2001 (Harvard University Press).

ELIZABETH LUNBECK is associate professor of history at Princeton University. She is the author of *The Psychiatric Persuasion: Knowledge, Gender, and Power in Modern America* (Princeton University Press, 1994), which was awarded the History of Women in Science Prize, and the editor of several books on the history of the human sciences. She is currently working on the history of psychoanalysis and its early practice in the United States.

PAMELA E. MACK is a professor in the Department of History at Clemson University, where she teaches a variety of courses in history of technology, including one she developed for first-year engineering students at the request of the College of Engineering. She is the author of *Viewing the Earth: The Social Construction of the "Landsat" Satellite System* (MIT Press, 1990) and editor of *From Engineering Science to Big Science: The NACA/NASA Collier Trophy Research Project Win-*

ners, NASA SP-4219 (Government Printing Office, 1998). She also publishes on the history of women in science and engineering.

MICHAEL S. MAHONEY teaches history of science and technology in the Program in History of Science at Princeton University and is the author of a variety of studies on mathematics and the mathematical sciences from antiquity through the seventeenth century. His current research deals with the history of computing, with a special focus on software and on the development of theoretical computer science and software engineering as new technical disciplines.

EMILY MARTIN is professor of anthropology at Princeton University. Author of *The Woman in the Body: A Cultural Analysis of Reproduction* (Beacon Press, 1987) and *Flexible Bodies: Tracking Immunity in America from the Days of Polio to the Age of AIDS* (Beacon Press, 1994), her present work addresses theories of normalization and the evolving constitution of selfhood in contemporary U.S. society.

RUTH OLDENZIEL received her Ph.D. in American history from Yale University and is currently associate professor at the University of Amsterdam, the Netherlands. She writes on cars, engineering, and household technology. She is the author of *Making Technology Masculine: Men, Women, and Modern Machines in America, 1870–1945* (Amsterdam University Press, 1999) and coedited the special issue on gender in *Technology and Culture* 38 (1997), *Schoon genoeg* (SUN, 1998), and *Crossing Boundaries, Building Bridges: Comparing the History of Women Engineers* (Harwood Academic Publishers, 2000).

NELLY OUDSHOORN is professor of gender and technology in the Department of Philosophy of Science and Technology at the University of Twente. Her research interests include gender, science, and technology. She is author of *Beyond the Natural Body: An Archaeology of Sex Hormones* (Routledge, 1994) and coeditor of *Bodies of Technology: Women's Involvement with Reproductive Medicine* (Ohio State University Press, 2000).

CARROLL PURSELL took his doctorate at the University of California, Berkeley. Since 1988 he has been at Case Western Reserve University, where he is the Adeline Barry Davee Distinguished Professor and chair of the Department of History. He has been president of the Society for the History of Technology and is president of the International Committee for the History of Technology. He is the author of, most recently, *The Machine in America: A Social History of Technology* (Johns Hopkins University Press, 1995).

KAREN A. RADER teaches science, technology, and society at Sarah Lawrence College in New York. Her work focuses on the history of twentieth-century U.S. life sciences, particularly the relations between laboratory work and political, social, and ethical issues in American culture. She is currently completing a history of the development and use of laboratory mice from 1900 to 1960.

LONDA SCHIEBINGER is the Edwin E. Sparks Professor of the History of Science at Pennsylvania State University. She is the author of *The Mind Has No Sex? Women in the Origins of Modern Science* (Harvard University Press, 1989), the prize-winning *Nature's Body: Gender in the Making of Modern Science* (Beacon Press, 1993), and *Has Feminism Changed Science?* (Harvard University Press, 1999) and is the editor of *Feminism and the Body* (Oxford University Press, 2000). She was the first woman historian to win the senior international Alexander von Humboldt Forschungspreis and was a research fellow at the Berlin Max Planck Institut für Wissenschaftsgeschichte during the academic year 1999–2000. Her current research explores gender in the European voyages of scientific discovery.

ALISON WYLIE teaches philosophy at Washington University in St. Louis. She is the author of *Thinking from Things: Essays in the Philosophy of Archaeology* (University of California Press, forthcoming). Her articles appear in *Osiris, Perspectives on Science,* and *American Antiquity* and in collections such as *Breaking Anonymity* (ed. The Chilly Collective, Wilfrid Laurier University Press, 1995), *The Disunity of Science* (ed. Peter Galison and David J. Stump, Stanford University Press, 1996), and *Women in Human Evolution* (ed. Lori D. Hager, Routledge, 1997).

INDEX